Walking Edinburgh's
LOST RAILWAYS

City & Suburban Explorations

ROBIN HOWIE & JOHN MCGREGOR

First published in 2026 by Whittles Publishing, an imprint of Porto Press.

The editors have made every effort to ensure the accuracy of information contained in this publication, but assume no responsibility for any errors, inaccuracies, inconsistencies and omissions. Likewise, every effort has been made to contact copyright holders. If any copyright material has been reproduced unwittingly and without permission the Publisher will gladly receive information enabling them to rectify any error or omission in subsequent editions.

British Library Cataloguing in Publication Data
A CIP record for this book is available from the British Library

ISBN: 978-1-84995-602-4

Cover and text design by Raspberry Creative Type, Edinburgh

Printed and bound in Great Britain by CPI

To order please go to our website www.portopress.com or contact our distributor, BookSource, 50 Cambuslang Road, Clydesmill Industrial Estate, Glasgow G32 8NB. Telephone 0141 642 9192

Porto Press Ltd
3 Connaught Road
St Albans
AL3 5RX

www.portopress.com

Paper from responsible sources

CONTENTS

Acknowledgements — v

Introduction — vii

Edinburgh's railways – a brief history — 1

North by Granton — 19

Waverley Station and Sir Walter Scott — 23

Walk Notes — 28

Walk 1: Waverley Station via Scotland Street to Granton Harbour 3 miles — 29

Walk 2: Waverley via Abbeyhill to Trinity Junction 2½ miles — 39

Walk 3: Roseburn to Granton 3½ miles — 48

Walk 4: Crewe Junction to Leith North Station 3 miles — 57

Walk 5: Warriston Junction to Leith Citadel Station 1½ miles — 64

Walk 6: Newhaven Station to Leith East Goods Yard 3 miles — 71

Walk 7: Abbeyhill Station to Leith Central Station 1 mile — 80

Walk 8: Balgreen Halt to Corstorphine Station 1 mile — 89

Walk 9: Craigleith Junction to Barnton Station 2½ miles — 94

Walk 10: Balerno Branch 7 miles — 98

Water of Leith Visitor Centre to Colinton Station 1½ miles — 100

Colinton to Juniper Green Station 1¼ miles — 103

Juniper Green to Currie Station 1¼ miles — 106

Currie to Balerno Station 1½ miles — 108

Balerno to Ravelrig Junction 1¼ miles — 110

Walk 11: Edinburgh & Dalkeith Railway 2½ miles — 113

St Leonard's to Duddingston Park Road 2 miles — 115

Dalkeith to Glenesk Junction ½ mile — 118

Walk 12: Queensferry Junction, Ratho to Queensferry 5 miles — 120

Ratho to Kirkliston Station 1¾ miles — 122

Kirkliston to Queensferry Station 3½ miles — 125

Walk 13: Bangour Branch (Bangour Hospital Railway) 1½ miles — 131

Walk 14: The Sub – Craiglockhart to Newington 2½ miles — 138

The Future? — 145

Station openings and closures — 147

Caledonian Railway and constituents, LMS and BR — 147

North British Railway and constituents, LNER and BR — 148

Further Reading — 150

ACKNOWLEDGEMENTS

My sincere thanks go to the many friends who accompanied me on the walks and to those who, later, armed with draft scripts, researched the routes –

Judy and Goff Cantley, Derek Chambers, Morag Dunbar, Frances Dunn, Rhona Fraser, Bill Howie, Anne and Wilson Middleton, and Philip Rankin. Many walkers we met along the way had a mine of remembrances, such that nostalgia has become an important part of the book.

I am particularly grateful to my twin sister, Frances, for much of the historical research which included the multitude of Edinburgh records of her late husband, Dr Munro Dunn.

My wife, Margaret, generously understands the author's often eccentric interests and always gives her whole-hearted support.

Derek Chambers, a Fellow of the Institution of Civil Engineers, spent a 40-year career in the construction industry. From 2011 until his retirement in 2014, he was Corporate Social Responsibility Director with the Forth Crossing Bridge Constructors (FCBC) on the Queensferry Crossing construction project. Patiently guiding the authors through the construction complexities of the Forth Bridge and the Queensferry Crossing, his contribution to the Ratho to Queensferry Walk script was invaluable.

My thanks go to Tom Wright for his contribution to the Walk to Leith Central Station – a wealth of information about the immediate post-war period when football games throughout the country were watched by crowds in unprecedented numbers. As if that was not enough, Tom has given permission to include many photographs from his private collection, a treasure trove gathered over 30 years of trading as owner of the Old Edinburgh in Pictures framing shop. In his book, *Glimpses of Times Past: Leith*, he opens the door to Leith's past by exploring its rich heritage through his extensive collection of old photographs.

I also thank Douglas Lowe, Secretary of the Currie & District Local History Society, for permission to use the photograph of Balerno Passenger Station; Roman Winkler for sight of his collection of Suburban line old tickets; and Emma Peattie of West Lothian Council for providing the striking image of an ambulance train at Bangour.

Every effort has been made to correctly acknowledge the source of images but, if there are any errors or omissions, please inform the publisher with a view to correction in any future editions.

Robin Howie, June 2026

Thanks are due to more than a few individuals for advice and assistance with photographs – but especially to Ewan Crawford (RailScot), David King (North British Railway Study Group), Hamish Stevenson and John Yellowlees. The late Bill Lynn allowed unrestricted access to his unequalled collection of North British images; the late Stuart Sellar was equally generous with his own photographs. John Furnevel, David Panton, Bill Roberton and Don Shaw made their work available on the easiest of terms.

For the inevitable faults in my summary history, I am alone responsible, and I owe much to those who have written more thoroughly and authoritatively about Edinburgh's railways or otherwise corrected my misapprehensions – in particular, David Ross but also several members of the NBRSG. I'm very grateful to Donald Cattanach for sight of his extensive notes on the evolution of Edinburgh's principal railway station and the mythology surrounding the name 'Waverley'.

John McGregor, June 2026

INTRODUCTION

Within the City of Edinburgh there are many miles of 'dismantled railway' (the standard Ordnance Survey term), which have been transformed into smooth, gently-graded, tarmac routes – ideal for pedestrians, young families, joggers, runners and cyclists. The exceptions are the conclusion of Walk 10 (Balerno Branch), and parts of Walk 12 (Queensferry Junction, Ratho to Queensferry) which both reach beyond the city limits as generally understood. Also, a very short stretch of Walk 9 (Craigleith Junction to Barnton Station).

Whether in town or country, railway engineers set a ruling gradient and balanced excavation and 'fill'. This has left a legacy of cuttings and embankments, often surprisingly tucked away from city life and hence more intimate and mysterious. Every city is more or less an aggregation of 'villages' and no one knows every 'village'. *Real* villages, with a long history of their own, have been caught up in Edinburgh's urban expansion.

In 1981 Lothian Regional Council acquired various disused lines and other railway land, making possible the creation of cycleways-cum-footpaths. This programme began in 1983 and the former Powderhall freight line – once a passenger route to Leith and Granton via Abbeyhill and retained into the 21st century for refuse traffic – may become the latest addition. More might have been achieved had the closures of the 1960s been accompanied

Edinburgh's so-called Binliner Express at the Powderhall loop. *Furnevel*

A modern diesel-electric locomotive on Waste Transfer duty. *Roberton*

by comprehensive plans for reutilisation, which the mood of the times did not favour. There was instead piecemeal redevelopment for commercial and industrial purposes, and hit-and-miss housing schemes of variable quality. Although a more enlightened policy now prevails, these results are not easily undone.

Nevertheless, the maze of trackbeds has very largely been preserved. Covering a total of some 40 miles, and all on easy gradients, 13 walks are described one by one. The first examples start from Waverley Station – a recognisable point for visitors to the city. It is suggested that the walks are best tackled in the order given.

Where trackbeds have been lost to urbanisation and industrial development, especially through parts of old Leith, exploratory detours are described in detail. Allow for more time than the distance may suggest!

Five Ways

The Caledonian Company's line to Leith (1864), running eastward from Pilton, spanned the Edinburgh, Leith & Granton Company's earlier line (1842) from Scotland Street to Trinity. (First styled Edinburgh, Leith & Newhaven, the latter had completed their local system in 1846; and operation from Canal Street in the heart of the city commenced the following year.) The two railways intersected without connection. Edinburgh, Leith & Granton quickly

The remains of Trinity Junction, North British Railway – today's Five Ways. The overgrown trackbed of the Scotland Street-Granton line tends to the right. The track curving left marks the later route via Abbeyhill. Note the piers of the dismantled overbridge carrying the Caledonian line to Leith. *Stevenson, courtesy NBRSG*

Five Ways interchange for cyclists and walkers: the view is to the east, along the sometime Caledonian route, now the path to Trinity Park and Newhaven. *Robin Howie*

The Five Ways signpost. *John McGregor*

became Edinburgh, Perth & Dundee (1849) and Edinburgh, Perth & Dundee then became North British (1862). In order to achieve more practical access to the Forth harbours and eliminate the cable-worked tunnel between Scotland Street and Canal Street, the North British built a new line (1868) from Abbeyhill, which divided for Leith and Granton near Powderhall. The Granton arm joined the original Granton route line at Trinity Junction, and the Caledonian overbridge, standing at almost the same spot, was modified accordingly.

Edinburgh's transport past is more thoroughly treated in the opening section of the book. This note concerns one happy legacy of these now defunct lines. With the old Caledonian overbridge demolished and the formation lowered to the level of the once North British tracks, a general interchange for walkers and cyclists could come into being. From any direction of approach, there would be four other paths from which to choose and many combinations to contrive. The designation 'Five Ways' at once suggested itself.

EDINBURGH'S RAILWAYS — A BRIEF HISTORY

Edinburgh's topography sets the scene. From the Pentland Hills the land falls to the Firth of Forth, through several folds. The glaciated hollow now occupied by Princes Street Gardens (and the railway west from Waverley Station) once contained the Nor Loch (drained between 1813 and 1820). While the medieval town scarcely grew beyond the crag-and-tail which declines eastward from the Castle, the expanding built-up districts of the 18th and 19th centuries, including much of the classic New Town, took shape for the most part on north-facing slopes. It bears remembering that even in the early 20th century the city remained

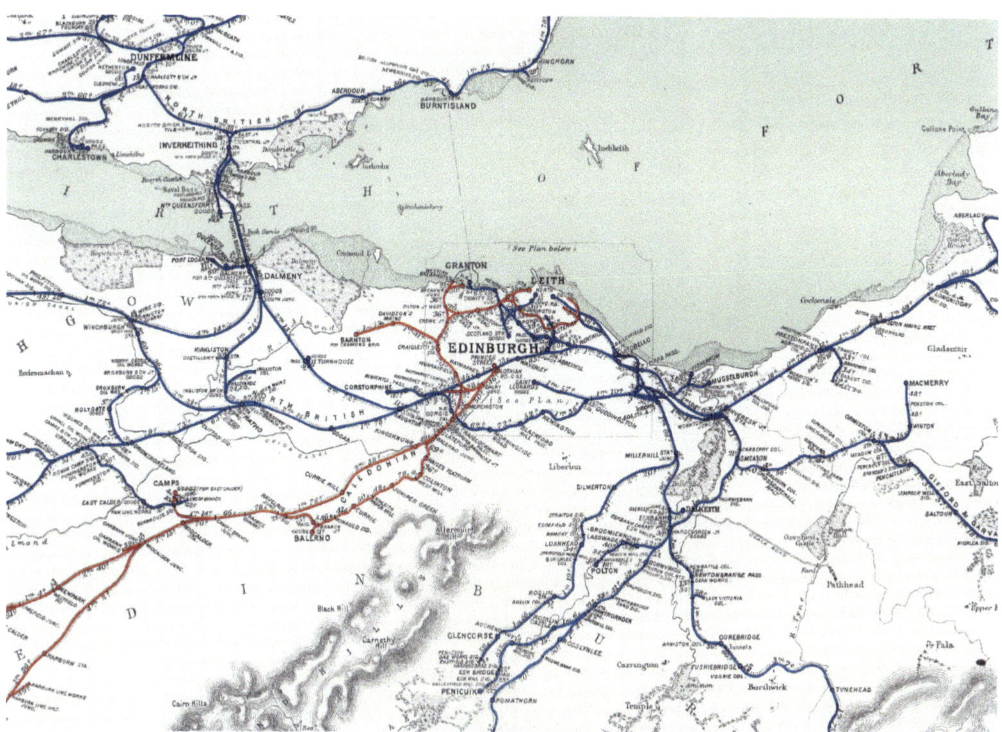

Caledonian Railway and North British Railway in and around Edinburgh. Reproduced with the permission of the National Library of Scotland.

compact. Leith, subordinate to Edinburgh until 1827 despite the importance of its harbour, had become thereafter an independent burgh; and the loss of this status, barely a 100 years later, caused much resentment.

Beginnings

Historians have identified an intermediate period in railway development – approximately the first three decades of the 19th century – when primitive wagon ways were already outmoded but an inter-city network, if not undreamed of, still lay in the future; when the limitations of horse-haulage and gravity working were already apparent, but steam traction had not yet convincingly proved itself. That railways could partner water transport was not in doubt. The Monkland lines across the Central Belt (1826–40) are a relatively late but clear case in point, feeding the Forth & Clyde Canal in the west and the Union Canal in the east; and a rail link to Leith docks from the Union Canal's Edinburgh basin (Port Hopetoun) was at least a possibility.

Like the Monkland lines, the Edinburgh & Dalkeith Railway (incorporated in 1826) was built to 'Scotch gauge' (4 feet 6 inches), 2½ inches less than the eventual Stephenson

The remains of South Leith station, Edinburgh & Dalkeith Railway. *NBRSG*

standard; and from 1831 it grew into a quite elaborate local system embracing the lower Esk Valley, South Leith and Fisherrow.

The company's passenger service proved popular; but coal traffic predominated, and St Leonard's, their base on the south-eastern flank of the Old Town, was never likely to become a station for the whole city. The line dropped steeply to Duddingston by a short tunnel behind Arthur's Seat – this section was at first cable-operated. Under its later owner, the Edinburgh & Dalkeith would be converted to standard gauge and equipped for locomotive operation (below). From 1847 regular passenger traffic ceased, and in 1859 the Baileyfield-Niddrie section would be abandoned. A detailed map from the 1870s suggests that the St Leonard's coal yard long displayed something of its original layout – an array of turnplates and short 'pigeon-hole' sidings[1].

The Edinburgh & Glasgow Railway was Scotland's first inter-city line. From c.1830, different routes were debated and that via Falkirk, unexacting but a little circuitous, finally found favour. There were shapely viaducts across the Almond and Avon valleys, while embankments, cuttings and tunnels on a lavish scale ensured generous curvature and imperceptible gradients – save at the Glasgow end, where a formidable incline in a noisome bore imposed enduring difficulties. (Working by cable was to continue until 1908, and Glasgow Queen Street to this day presents an operating challenge, despite electrification and bi-directional signalling.) Authorised in 1838, the line was completed in 1842, with the company's Edinburgh terminus at Haymarket, on the city's western edge, and their nearby locomotive depot 'out in the country'. West from Haymarket, there were stations at Saughton ('Corstorphine'), Gogar and Ratho; and the Edinburgh & Bathgate Railway, diverging just beyond Ratho and completed in 1849, became at once a subsidiary – it would eventually be extended to Airdrie and Glasgow. (But Lenzie, some six miles out of Queen Street, which speedily grew into a Glasgow dormitory, had no Edinburgh counterpart.) Though closely associated with the Edinburgh & Glasgow, the Monkland Company (their lines widened to standard gauge) would remain formally independent into the 1860s. From 1848 the Scottish Central Railway gave connection with Stirling, Perth and the north, at first by reversal at Greenhill, west of Falkirk, but subsequently via Polmont and Larbert – whereupon the Edinburgh & Glasgow claimed the right

Edinburgh & Glasgow Railway insignia.

---

1 D. Spaven, *Railway Atlas of Scotland*, 2015

St Margaret's, Meadowbank – during 2022 the site of one of the turntables at the former locomotive depot was excavated and a record made. *Roberton*

to share in northern traffic, while conceding a balancing right of access to Haymarket for Scottish Central trains.

In 1844 the North British Company obtained parliamentary sanction for their Edinburgh-Dunbar-Berwick main line, beginning at the old North Bridge, with a branch from Longniddry to Haddington. Construction was rapid and traffic began just two years later. The North British locomotive depot was established at St Margaret's, on the northeastern side of Edinburgh's Holyrood Park.

Acquired in 1845, the Edinburgh & Dalkeith Railway provided the stem for advance into the Borders. The Edinburgh & Hawick Railway, a North British subsidiary, was authorised that same year and commenced at Dalhousie. It would grow into the Waverley Route, which reached Carlisle in 1862. A modest branch to Musselburgh, diverging from the Fisherrow line (above), was completed in 1847.

The North Berwick branch, authorised in 1846 and opened in 1850, was part speculative, part strategic (so as to keep East Lothian exclusively North British), but the hoped-for 'residential' business developed only slowly.

Opened in part in 1842, the Edinburgh, Leith & Newhaven Railway became from 1844 the Edinburgh, Leith & Granton. The line fell sharply from its Canal Street terminus (later the site

Musselburgh station and yard in their last years, when goods traffic was still sufficient to employ a diesel shunter and the passenger service had been entrusted to DMUs. *NBRSG*

Canal Street station, Edinburgh, Perth & Dundee Railway, at the angle of Princes Street and the then Waverley Bridge; the line descended in tunnel to Scotland Street. The Scott Monument (1844) and Royal Scottish Academy (1826) appear in the background.

of Waverley Market) to Scotland Street, by a tunnel some two-thirds of a mile long, where cable operation was again prescribed. ('Canal Street', an incongruous designation but a reminder of the abortive plan to carry the Union Canal all the way to Leith, was aligned approximately as Waverley Station's north access-ramp runs today, ending below the old North Bridge.)

Construction was interrupted when pent-up underground springs burst into the headings, killing four workmen[2]. The directors briefly toyed with atmospheric traction, against the possibility of a rival promotion, paralleling Leith Walk, which would have employed this passing technological fad. Their ambitions were not merely parochial – because Granton soon became the city's railhead for traffic across the Forth, whence the Edinburgh & Northern Railway, opened in 1848, gave connection to Tayport (for Dundee) and Perth (via Ladybank and Bridge of Earn). A year later the two companies amalgamated to form the Edinburgh, Perth & Dundee Railway. In 1850 a pioneering train ferry was introduced between Granton and Burntisland, and similar provision across the Firth of Tay followed. As a rule, only goods and minerals were carried; passengers used conventional, better-appointed vessels. Thereafter the 'Ferry Route', for all its limitations, competed in some measure for long-distance traffic with the unbroken Edinburgh & Glasgow-cum-Scottish Central route via Larbert and Stirling.

2 *Caledonian Mercury*, 2 December 1844, courtesy R. Copson and North British Railway Study Group

Kingsknowe signal box and level crossing, on the former Caledonian main line into Edinburgh. Note the large finial. *Roberton*

Edinburgh's so-called "General Station" grew between the Waverley Bridge and the old North Bridge, while Canal Street stood apart. Note the original Mound tunnel. The National Gallery (1859) is under construction.

In 1846 the Edinburgh & Glasgow was extended eastwards, in tunnel and then (controversially) through Princes Street Gardens. The extension pierced the Mound (by a lesser tunnel) to join end-on with the North British. A tightly curved connection to the Edinburgh, Perth & Dundee Company's Canal Street came later; and the net result was something of a Hautbahnhof, or Union Station, shared by all-comers. This grew piecemeal between the North Bridge and the Waverley Bridge. At first known as the 'General Station' (though this was never official), in popular usage it became 'Waverley', which title was eventually adopted[3].

But missing from this picture is the Caledonian Company, who aspired to possess the sole – or, at any rate, the most important – Anglo-Scottish trunk railway and in due course to command all southern Scotland. Approved in 1845 and opened in 1848, the Caledonian main line reached north from Carlisle, dividing at Carstairs for Edinburgh and Glasgow; and it included a further arm to Greenhill, making end-on connection with the Scottish Central. From its summit at Cobbinshaw (near the principal reservoir for the Union Canal), the Carstairs–Edinburgh line traced the flanks of the Pentlands, ending in make-shift premises on Lothian Road.

The Caledonian's showpiece terminus – the foundation stone was laid in 1847 – did not materialise, and the short advance to their more commodious but undistinguished Princes Street Station, fronting the West End, came only in 1870. Though the city council favoured a truly *general* station at Waverley Bridge, accommodating the Caledonian too, inter-company conflict proved insurmountable. North British and Caledonian were rivals: both had large

---

3 See 'Waverley Station and Sir Walter Scott'

Relics of the short-lived Edinburgh & Northern Railway long survived the amalgamations which by 1862 had brought the North British Company into Fife. Examples of Scottish Central signage likewise endured under Caledonian ownership. *NBRSG*

ambitions within Scotland, as their respective titles implied; both handled cross-Border traffic. The so-called West Coast Alliance, with the Caledonian as Scottish partner, was in the making from 1846, while by 1850 the competing East Coast partnership, which included the North British, had taken firm shape.

The 'Great Amalgamations' and an enduring rivalry

On into the early 1860s fragile treaties of alliance, in various permutations, were made, unmade and remade by Edinburgh's four railway companies. However, control of the Central Belt was key, and the long-term trend pointed to victory for the Caledonian, who had the deepest purse and twice forced the Edinburgh & Glasgow into near dependency. (Their attrition tactics included an alternative, loss-making, city-to-city service via Carstairs.) The North British, though dominant within the Edinburgh–Berwick–Carlisle triangle, risked subordination to their English neighbour, the North Eastern Company, whose territory eventually stretched from York to the Border. Or, as some observers expected, Caledonian and North Eastern might between them have squeezed the North British into relative insignificance. Nevertheless, matters in the end turned out otherwise. In 1862 the North British absorbed the Edinburgh Perth & Dundee[4]; in 1865 they acquired the Edinburgh & Glasgow, together with the Monkland system. By way of balance, Parliament permitted their enemy to acquire both the Scottish Central Railway (1865) and the Scottish North Eastern Railway (1866) – which extended the

---

1 Strictly speaking, both companies were dissolved and reincorporated under the North British title.

Caledonian system to Aberdeen, though with safeguards for North British traffic beyond the Tay. In taking over the Edinburgh & Glasgow, the North British secured the right to send their northern traffic via Stirling, while the Caledonian inherited the Scottish Central's running powers from Larbert into Edinburgh[5]. It proved definitive for Scottish railway history. Reconstituted, the North British and the Caledonian were to endure until 1922 – together with the Highland Railway, the Glasgow & South Western and the Aberdeen-based Great North of Scotland (all three were of lesser scale). North British and Caledonian had emerged from these amalgamations roughly equal, and by the early 20th century the one would command just over, the other just under, 1,000 route-miles. However, the Caledonian possessed the more coherent empire. Though eventually bound together by the Forth and Tay Bridges ('Bridges Route' finally superseded 'Ferry Route' in 1890), the North British network remained quite recognisably a patchwork.

The Edinburgh, Leith & Granton had anticipated an early Caledonian bid for access to the Firth of Forth and might have joined the Edinburgh & Glasgow in opposition; but

---

5 Haymarket West Junction, giving access to Caledonian Princes Street from the Larbert direction, was not in use until 1876.

serious Caledonian efforts to expand in and around Edinburgh were to come somewhat later, beginning with their Granton branch, authorised in 1857 and opened in 1861. This diverged at Dalry Junction, to span the Edinburgh & Glasgow (just west of Haymarket), while a connection from Slateford permitted through running from Carstairs, Glasgow and Carlisle.

No doubt the Caledonian looked to intimidate the Edinburgh, Perth & Dundee Company (above), who had not yet committed to union with the North British. Extension to Leith followed in 1864, from a triangular junction at Crew-Pilton and initially for goods traffic. A passenger service was introduced five years later. This challenged the by now enlarged North British Company, whose response would be a new Leith-and-Granton route via Abbeyhill (1868). Though circuitous, it greatly improved on the original line of 1842 (above), and a connection in the Portobello direction, through Piershill, was included. The Scotland Street incline and tunnel thus became redundant. Canal Street, for a time restyled 'Princes Street' (a petty swipe at the Caledonian enemy?) would more or less disappear, thanks to the ongoing, piecemeal expansion of the east-west aligned 'General Station'. Curiously, the new line from Abbeyhill to Trinity Junction cut the original Scotland Street–Leith tracks on the level and at right angles (in railway terminology, a flat crossing). Two spurs gave direct access to Leith – Bonnington South Junction–Bonnington East Junction (from the Abbeyhill direction) and Bonnington North Junction–Bonnington East Junction (from the Granton direction). Adding to the tangle of lines by now established between central Edinburgh and the shores of the Forth, the early 20th century was to see a renewed and wastefully expensive battle for Leith (below); but the districts in question are today altogether devoid of working railways, save for the little-used tracks between Portobello and Seafield.

From the 1860s to the 1920s

After absorbing the Edinburgh, Perth & Dundee, the North British looked to achieve a through connection across the Forth. A low-set viaduct from Blackness to Charleston, more than two miles in length but within existing engineering capabilities, was seriously investigated – on the basis that the Edinburgh & Glasgow Company would share the costs. However, the resulting report was discouraging, while Bo'ness, Grangemouth and Alloa resolutely resisted the threat of shipping restrictions in the upper firth. The North British next obtained powers to acquire the Queensferry passage, only to have Parliament reject their proposed extension from Granton, whereby these powers could be made effective. A protracted quarrel loomed, given that an Edinburgh & Glasgow branch from Ratho to South Queensferry via Kirkliston had been approved instead; but negotiations for amalgamation, already begun, were to prove successful, and the authorised line opened in 1868, under North British ownership. It was subsequently extended to Port Edgar. Matters then rested until the completion of the Forth Bridge (1890) and the connecting lines from Saughton and Winchburgh. Thereafter passenger trains from Ratho either terminated at the new Dalmeny

Such signs, citing railway company by-laws, were once common-place – at stations, goods yards and locomotive depots – and could readily be found around Edinburgh. *NBRSG*

station, or ran on across the Forth to Dunfermline, and the South Queensferry branch became something of a backwater, though goods traffic remained worthwhile into the 1950s. (The branch was temporarily re-energised by the requirements of the Royal Navy during the 1914–18 War.) The North British introduced (1891) an intensive service (by then standards) between Edinburgh and Fife; and the original two-platform Saughton station (also known as 'Corstorphine' and for a time restyled 'Saughton Junction') would become four-platform (below). Turnhouse was the only station added between Saughton and Dalmeny.

South Gyle and Edinburgh Gateway, like Edinburgh Park on the Glasgow route, are latter-day additions, while the Almond Chord, today's projected relief line from Winchburgh by Kirkliston to Saughton, has yet to become a reality.

In 1865 the Caledonian Company endorsed a branch to Balerno. Construction was postponed when the funds immediately available were diverted to consolidating their union with the Scottish Central and Scottish North Eastern (above), the better to confront their great rival. The continuing enmity of Caledonian and North British, both now enlarged, was exemplified in the Caledonian's Midcalder–Shotts–Holytown cut-off, completed by 1869. This created an Edinburgh–Glasgow route which no longer detoured by Carstairs; and thereafter the Shotts line, a little shorter (though distinctly hillier) than the ex-Edinburgh & Glasgow route via Falkirk, offered the North British enduring competition. The two foes *might* have made peace and come together, crowning all the earlier amalgamations of 1862–6, and in 1871–2 a settlement was very nearly achieved – with portentous implications for Scotland's railway system. The combined company need not have hastened to conquer Forth

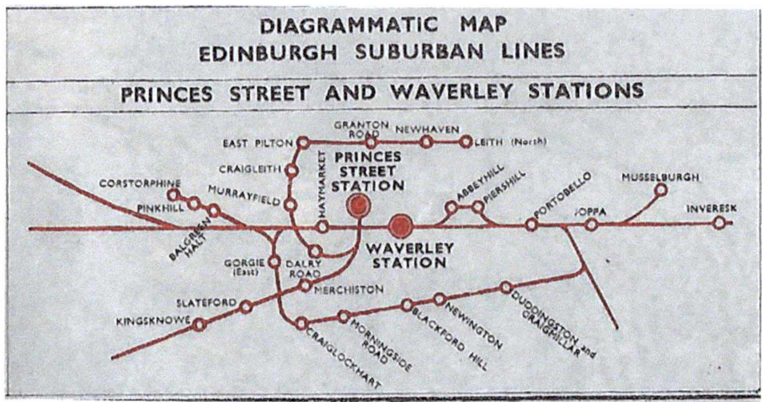

From British Railways (Scottish Region) summer timetable 1957, when several Edinburgh suburban services were still precariously in being.

and Tay; and development elsewhere, notably in both Edinburgh and Glasgow besides the poorly served western Highlands, would surely have taken a different course. But the moment passed, and 20 more years of costly warfare ensued, until the so-called Peace Agreement of 1891[6] brought fragile détente and grudging co-existence. The North British position, with the great bridges which unified their system as yet unbuilt, was bolstered in the medium term when England's Midland Railway reached Carlisle. From 1876 Midland and North British formed a new Anglo–Scottish partnership. The latter's bargaining power within the older East Coast alliance was increased and the Waverley Route gained much-needed revenue. In any case, Edinburgh remained, railway-wise, a divided city. It bears mention, too, that the North Eastern Company's locomotives, stabled at Haymarket depot from 1865, had become not altogether welcome guests of the North British, taking the principal East Coast passenger trains, i.e. without change of engine at Berwick; however North British domestic traffic was unaffected.[7]

Opened in 1884, the Edinburgh Suburban & South Side Junction Railway (soon christened 'The Sub') was a North British client and tapped the growing population of all the districts between Liberton and Craiglockhart[8]. At Duddingston 'The Sub' diverged from the St Leonard's arm of the ex-Edinburgh & Dalkeith, and there was connection eastwards with the Edinburgh–Berwick main line; at its other end 'The Sub' connected in both directions with the main line through Haymarket.

An east-west relief route thus came into being, which was the primary aim. Without it, the haphazardly expanded 'General Station' – when officially christened Waverley, it remained no less a make-shift – could not have coped for much longer. Professor Foxwell,

---

6 Properly 'New Lines Agreement', this defined 'exclusive districts' and 'shared districts'.
7 See also 'Waverley Station and Sir Walter Scott'
8 See 'Walk 14: The Sub'

a trenchant critic, poked fun at the 'narrow village platforms' which the North British 'reserved for important expresses'[9]. Meanwhile, the Caledonian's Balerno line (above), completed in 1874, had at once assumed an outer-suburban character – although the goods traffic of the several mills along the Water of Leith, not heavy passenger use, had been the chief consideration when the branch was conceived. It was for freight business that it looped back to the Caledonian main line at Ravelrig. (A few passenger trains continued beyond Balerno to West Calder or Addiewell.) Because curvature was severe, short, rigid-wheelbase passenger rolling stock continued to be employed on into the 1950s.

These additions notwithstanding, even by late Victorian standards Edinburgh's built-up area remained tightly defined, and the several North British branches built south and east of the city between the 1850s and the 1870s – to Bonnyrig (1855), Smeaton (1866), Lasswade and Polton (1867), Penicuik (1872), MacMerry (1872), Gilmerton and Loanhead (1873), and Roslin and Glencorse (1874) – would never achieve what seems in hindsight their outer suburban potential.

The emphasis was on mineral traffic and these lines saw no multiplication of intermediate stations for 'residential' and commuter business. Well-to-do 'residenters' (sic) for the most

9 Foxwell & Farrer, *Express Trains English and Foreign*, 1889 (Ian Allan Reprint)

'Swindon' DMU and steam 'special'* meet (c.1960) at Rosewell & Hawthornden – a station within easy commuting distance of Edinburgh. But until the late 20th century, the area south of the city saw little outer-suburban encroachment. *Sellar*

* Headed by the restored North British locomotive 'Glen Douglas', to be found today at Glasgow's Riverside Museum.

part looked further afield – say, to West Linton, behind the Pentlands (served by the Leadburn-Dolphinton branch)) or to Aberlady and Gullane in East Lothian (which later obtained their own branch). The 'residenters' contributed considerably to North British income but scarcely belong in this Edinburgh overview.

Two turn-of-the-century suburban schemes, on the then western margin of the city, were the Caledonian Company's branch from Craigleith to Barnton (1894) and the North British branch from their main line beyond Haymarket (Balgreen) to Corstorphine (1902). The Caledonian also undertook an ambitious, expensive and (in North British eyes) provocative extension to Leith East (1903). Though there was goods traffic to be won, further passenger provision in that quarter was little wanted and the planned stations never opened – which makes their rival's response seem an over-reaction. The North British obtained powers for a short branch from the Abbeyhill–Trinity line to a new station at the foot of Leith Walk. Named Leith Central, this imposing terminus resembled the Caledonian's new Princes Street (below) more than a little. A spur (Lochend North Junction–Lochend South Junction) gave connection eastward via the existing Abbeyhill–Piershill loop. In consequence there was access to 'The Sub' on Edinburgh's south side both through Waverley and through Portobello. From then on, fewer 'Sub' services traced a true circle, while others began or ended at Leith Central. North British secondary services to Glasgow were also accommodated there – westbound, these trains now ran via Abbeyhill, calling at Waverley and Haymarket, before taking the Bathgate route. In the opposite direction, arrangements were similar, thanks to the Glasgow City & District Railway, a North British dependent of 1884; trains bound for Bathgate, Edinburgh and Leith Central ran from Hyndland, in Glasgow's western suburbs, via Queen Street Low Level and Airdrie.

The North British decision to spend so lavishly was influenced in part by the increasing flow of Baltic migrants into Leith, whence many journeyed to the Clyde and took passage to the United States; but the flow would end abruptly on the outbreak of war in 1914. More important were old fears, now reignited, that the Caledonian meant to reach eastwards through Portobello, invading the Midlothian coal field. And rumours had grown of yet bolder designs – a completed Caledonian 'circle' from South Leith back to Caledonian Princes Street, tunnelling under Calton Hill and across the New Town[10], and perhaps

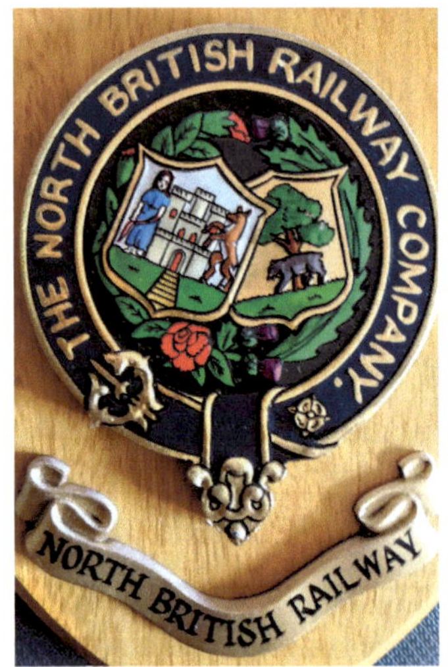

The North British 'Castle and Bear' heraldic device. A version in mosaic marked the four corners of Waverley's 1890s booking hall (today's general waiting area).

---

10 By shallow cut-and-cover, the full length of George Street or Princes Street, cf. the Caledonian-owned Glasgow Central Railway, along that city's Argyle Street.

also a supplementary outer circle, by new construction from Barnton to Balerno. Nevertheless, in popular estimation the far from faultless North British remained 'Edinburgh's railway', and the North British Company repaid the compliment by retaining throughout their whole existence their original armorial device representing Edinburgh and Berwick. The Caledonian, for all their efforts, never entirely lost the colour of an interloper from the West.

North British traffic, both cross-Border and within Scotland, had grown even more after 1890, with the 'Bridges Route' in being; and between 1894 and 1898 Waverley would be comprehensively rebuilt – in conjunction with four-tracking from Abbeyhill Junction to Saughton Junction (1893–7) and construction of the city's modern North Bridge. Meanwhile the shabby Caledonian terminus of 1870, which suffered severe fire damage in 1890, was replaced in much grander, well-proportioned style (1894). Though the

Edinburgh Princes Street, Hogmanay 1947

A handsome clock commanded the well-proportioned concourse of the Caledonian Company's rebuilt Princes Street terminus.

idea foundered almost at once, during the doubling of the North British mainline tracks across the city it was suggested that interchange platforms be inserted in the old and new Mound tunnel[11], with a stairway to remodelled Princes Street overhead.

There were two more latter-day projects, on either side of Edinburgh. Worked by the North British, the Bangour Asylum Railway opened in 1905; it ran from Uphall, on the Bathgate line, to Bangour Village Hospital. Planned before 1914 but opened under wartime 'Control' (below), the Lothian Coal Lines linked Leith (Seafield), Portobello, Niddrie and Monktonhall – which created additional capacity for local mineral and freight movements but did not impact directly on passenger traffic. The coal owners of Midlothian had long pressed the North British for some such project, to obviate congestion and delay; they had threatened, more than once, to turn to the Caledonian, which partly explains the latter's enthusiasm for their ill-advised extension to Leith East.

11 East from Waverley, a duplicate Calton tunnel was required too. The existing Mound tunnel was flanked by two new single-track bores.

Into the 1960s, lengthy mineral trains both loaded and empty, often hauled by veteran ex-North British locomotives, still dominated the Portobello-Seafield tracks of what had begun as the Leith arm of the Edinburgh & Dalkeith Railway. *Stevenson, courtesy NBRSG*

Decline

Government direction 'Control' of Britain's railways during 1914–19 might have led to post-war public ownership. Rationalisation and retrenchment, already looming before 1914, could no longer be postponed and the strategy eventually chosen (after Nationalisation had been ruled out) was 'De-Control', which restored private ownership, followed by Grouping, which created four regional blocks. The necessary amalgamations were thrashed out during 1920–2, leaving the five Scottish companies split between the London Midland & Scottish and the London & North Eastern Groups. Caledonian became 'LMSR' and North British 'LNER', and the division of Edinburgh's railways thus continued. (The option of an all-Scotland Group was rejected.)

The railway companies generally had experienced the inexorable rise of tram competition before 1914; but in Edinburgh this was tempered by the constraints of a cumbersome cable system – dating from 1888 and superseding the city's several horse tramways. Electrification waited until Edinburgh's union with Leith, where the trams were already 'modern'. Between the wars, the motor bus would become an even more serious competitor, with little scope for a railway fight-back. One LMS riposte, plucky though scarcely significant in the bigger scheme of things, was the creation of East Pilton station and House o'Hill halt, on the ex-Caledonian lines out of Princes Street – the one served by Leith trains, the other by Barnton trains. Ageing stations and reduced maintenance, in the name of economy, emphasised the smoke and grime inseparable from steam traction; and the Second World War would make exceptional demands on the railway network, which emerged even more overstretched and run-down. Emergency once again presaged outright Nationalisation, this time made certain by the Labour Party's victory in the general election of 1945. Though high hopes were invested in 'British Railways, Scottish Region' (in being from 1948), the story of the ensuing 30 years appears today as one of managed retreat, bedevilled by government short-termism. How this might have turned out better has become a question for historians to debate; but there is a consensus that management by ponderous public corporation reinforced the conservatism of an industry where tradition ruled. On into the 1960s, timetables, working practices and public relations all suggested that LMS and LNER, if not Caledonian and North British, were still in ghostly existence; and by 1970 closures, whether 'pre-Beeching' or 'post-Beeching'[12], had so reduced Edinburgh's local railways as to preclude revival on the scale which Glasgow was to enjoy in the more enlightened 1980s and 1990s.

---

12 Dr Richard Beeching's radical Report, 'The Reshaping of British Railways' appeared in 1963.

Imposing but under-used, Leith Central was something of a North British Company 'folly' and ended as an early DMU depot. The 'Swindon' DMUs (an example of which can be seen here) introduced in 1956 would sustain the Edinburgh-Glasgow intercity service until 1971. *NBRSG.*

The trimming of Edinburgh's suburban network between the 1940s and the 1960s is detailed in 'Station Openings and Closures' at the end of this book. Hitherto station closures were few – it hardly counts that the Bangour Asylum Railway, well outside the city, was abandoned in 1921. The so-called modernisation programme, abruptly adopted from the mid-1950s, meant the introduction of diesel multiple units (DMUs), which were expected to revive the fortunes of both inner-urban lines and rural branches. These locomotive-less, standardised trains (no 'run round' at journey's end, no shunting to add or subtract vehicles, no lengthy locomotive servicing) made for more flexible, more generous workings – for example, through trains between Corstorphine and North Berwick and additional services for local stations as far as Rosewell & Hawthornden, on the line to Peebles; but the boost to passenger numbers proved short-lived. Dieselisation, besides condemning Britain's latest steam engines to early and wasteful withdrawal (several Standard classes had been commissioned in large numbers after 1948), entailed much hasty improvisation in respect of stabling and maintenance. It makes a strange postscript to this short account, set largely in the era of steam and inter-company competition, that Scottish Region hit on the expedient of turning grandiose, always under-used, Leith Central, where passenger traffic had ceased in 1952, into a depot for their first-generation DMUs.

JUNCTIONS IN THE EASTER ROAD AREA

1. 1868 Abbeyhill, on Edinburgh-Berwick main line – for Leith/Granton, replacing route from Canal Street

2. 1868 Easter Road, on Leith/Granton line – to/from Piershill and lines eastward (Leith/Granton traffic and ultimately Powderhall waste traffic)

3. 1888 London Road, on Leith/Granton line, just beyond Abbeyhill station

4. 1888 Lochend, on Easter Road-Piershill line. The spur from 3 to 4 created a loop, Abbeyhill-Piershill, off the main line

5. 1903 junction for Leith Central added at London Road

6. 1903 Lochend North, on Leith Central line – for Leith Central traffic to/from Piershill and lines eastward. Lochend, where the spur from Lochend North came in, was renamed Lochend South; it reverted to Lochend in 1972 after all Leith Central traffic had ceased, making Lochend North redundant.

NORTH BY GRANTON

The pattern of passenger services developed by the Edinburgh, Perth & Dundee Company, successor to the Edinburgh, Leith & Granton, continued under North British ownership after 1862.

Though by later standards long-distance travel via the Forth and Tay Ferries was protracted and wearisome, the intervals allowed for rail/steamer transfer or vice versa at Granton, Burntisland, Tayport and Broughty Ferry, and likewise for shunting at Ladybank,

Burntisland station before the building of the Forth Bridge, where passengers northbound by the Granton ferry took their second train. The route initially included a second ferry and a third train. *Lynn*

seem extraordinarily brisk. The more generous allowance between Dundee[13] and Broughty is explained by the need for reversal to and from the short Ferry branch. Until the completion of the Forth Bridge nearly thirty years later, the unbroken route to the North by Larbert, Stirling and Perth (in the Caledonian Company's hands from 1866) enjoyed an obvious advantage and through Anglo–Scottish East Coast coaches for Aberdeen and Inverness necessarily ran via Larbert. (The Highland Railway's Tay-and-Spey route from Perth to Inverness dated from 1863.)

The Tay Bridge (1878), made possible a through Burntisland-Aberdeen journey via Dundee – but the slender structure collapsed under a train in 1879, during a December gale. The second Tay viaduct was completed in 1887, the Forth Bridge three years later.

Edinburgh and the North via the Forth and Tay ferries and Perth/Dundee (summarised)
(Based on the timetable for May 1863 reproduced by the North British Railway Study Group)

	A	A	C	C	D	D	C	C	
Princes Street (sic) d.		06.15		09.40		14.30		18.15	
Scotland Street d.		06.21		09.46		14.36		18.21	
Granton a.		06.33		09.58		14.48		18.33	
d.		06.40		10.05		14.55		18.40	F
Burntisland a.		07.08		10.33		15.23		19.08	F
d.		07.15		10.40		15.30		19.15	
Ladybank a.		08.20		11.40		16.28		20.20	
Ladybank d.	08.22	08.24	11.42	11.44	16.30	16.32	20.22	20.24	
Tayport a.		09.05		12.25		17.11		21.05	
d.		09.10		12.30		17.15		21.10	F
Broughty a.		09.18		12.38		17.23		21.18	F
d.		09.25		12.45		17.28		21.25	
Dundee [13] a.		0942		13.02		17.48		21.42	
Perth a.	09.05		12.25		17.13		21.10		
	(1)		(2)				(3)		

(1) Connection for Aberdeen via Forfar, arrive 12.35

(2) Connection for Aberdeen via Forfar, arrive 16.10

(3) Connection for Aberdeen via Forfar, arrive 03.00 next day.

---

13 At this date the North British used Dundee's Dock Street Station, built originally for the Dundee & Arbroath Railway (1838). It later became Dundee East.

	A	A	B	C	D	D	C	C	
Perth d.	06.30		10.40		14.50		18.35		
Dundee d.		05.50		09.45		14.10		17.55	
Broughty a.		06.05		10.00		14.25		18.10	
d.		06.10		10.05		14.30		18.15	F
Tayport a.		06.18		10.13		14.38		18.23	F
d.		06.25		10.20		14.45		18.30	
Ladybank a.	07.12	07.13	11.10	11.08	15.30	15.28	19.15	19.13	
d.		07.16		11.12		15.33		19.18	
Burntisland a.		08.22		12.01		16.31		20.20	
d.		08.30		12.09		16.39		20.28	F
Granton a.		08.58		12.37		17.07		20.56	F
d.		09.05		12.44		17.14		21.03	
Scotland Street a.		09.20		12.57		17.27		21.15	
Princes Street (sic) a.		09.30		13.05		17.35		21.25	
				(1)		(2)			

(1) Connection from Aberdeen via Forfar, depart 06.00
(2) Connection from Aberdeen via Forfar, depart 09.15

A 1st, 2nd and 4th Class passengers
B 1st, 2nd and 3rd Class passengers, non-stop Perth to Ladybank; preceded by 10.10 all-stations Perth–Ladybank, also with through coaches for Burntisland.
C 1st, 2nd and 3rd Class passengers
D Mail train, 1st, 2nd and 3rd Class passengers
F Ferry

Edinburgh–Leith–Granton

Pending completion (1868) of the longer but more easily operated line via Abbeyhill, the North British continued the well-established and relatively intensive service to Leith and Granton by the Scotland Street tunnel; and there were additional trains, or through coaches, between Leith and Granton, reversing at Scotland Street. The timings suggest that Scotland Street's regular routines – cable-to-locomotive, locomotive-to-cable and other re-marshalling – were performed with remarkable speed; and it seems that local passengers were not excluded from the more important trains, i.e. those giving long-distance connections via Burntisland. Ex-Edinburgh, Perth & Dundee Canal Street, which the North British liked to style Princes Street, might be described as 'attached' to Waverley Station. But the unofficial usage, 'General Station', loosely comprehended it, while the North British were content, inconsistently, with simple 'Edinburgh' for local purposes[14].

14 See also 'Waverley Station and Sir Walter Scott'

EDINBURGH, LEITH, AND GRANTON SECTION.

EDINBURGH TO GRANTON.

Trains dep.	a.m.	a.m.	a.m.	a.m.	a.m.	noon.	p.m.	p.m.	p.m.	p.m.	p.m.	p.m.	p.m.	p.m.	p.m.	p.m.	Sundays a.m.	Sundays p.m.
Edinburgh ... „	6 15	8 0	8 45	9 40	11 0	12 0	1 0	2 0	2 30	4 0	4 55	6 15	7 0	8 0	9 15	10 0	6 40	4 30
Scotland St. „	6 21	8 5	8 50	9 46	11 5	12 5	1 5	2 5	2 36	4 5	5 0	6 21	7 5	8 5	9 20	10 5	6 45	4 35
Trinity... „	6 28	8 10	8 55	9 53	11 20	12 10	1 10	2 10	2 43	4 10	5 5	6 28	7 10	8 10	9 25	10 10	6 50	4 40
Granton ... arr.	6 33	8 15	9 0	9 58	11 15	12 15	1 15	2 15	2 48	4 15	5 10	6 33	7 15	8 15	9 30	10 15	6 54	4 44

GRANTON TO EDINBURGH.

Trains dep.	a.m.	a.m.	a.m.	a.m.	a.m.	p.m.	p.m.	p.m.	p.m.	p.m.	p.m.	p.m.	p.m.	p.m	p.m	p.m.	p.m.	Sundays a.m.	Sundays a.m.	Sundays p.m.
Granton ... „	7 35	8 35	9* 5	10 20	11 20	12 20	12*44	1 20	2 20	3 20	4 20	5*14	6 40	7 20	8 20	9* 3	9 35	9 43	10 20	7 13
Trinity... „	7 38	8 38	9 10	10 23	11 23	12 23	12 49	1 23	2 23	3 23	4 23	5 19	6 43	7 23	8 23	9 8	9 38	9 50	10 23	7 20
Scotland St. arr.	7 42	8 42	9 20	10 27	11 27	12 27	12 57	1 27	2 27	3 27	4 27	5 27	6 47	7 27	8 27	9 15	9 42	10 0	10 30	7 30
Edinburgh ... „	7 50	8 50	9 30	10 35	11 35	12 35	1 5	1 35	2 35	3 35	4 35	5 35	6 55	7 35	8 35	9 25	9 50			

Passengers from Granton to Leith and Leith to Granton change carriages at Scotland Street.

LEITH TO GRANTON.

Trains dep.	a.m.	a.m.	a.m.	a.m.	a.m.	a.m.	p.m.	p.m.	p.m.	p.m.	p.m.	p.m.	p.m.	p.m.	p.m.	p.m.
Leith ... „	6 5	7 35	8 35	9 28	10 50	11 50	12 50	1 50	2 20	3 50	4 50	6 0	6 50	7 50	8 50	9 35
Bonnington „	6 8	7 38	8 38	9 31	10 53	11 53	12 53	1 53	2 23	3 53	4 53	6 3	6 53	7 53	8 53	9 38
Scotland St. „	6 21	8 5	8 50	9 46	11 5	12 5	1 5	2 5	2 36	4 5	5 0	6 21	7 5	8 5	9 20	10 5
Trinity... „	6 28	8 10	8 55	9 53	11 10	12 10	1 10	2 10	2 43	4 10	5 5	6 28	7 10	8 10	9 25	10 10
Granton ... arr.	6 33	8 15	9 0	9 58	11 15	12 15	1 15	2 15	2 48	4 15	5 10	6 33	7 15	8 15	9 30	10 15

GRANTON TO LEITH.

Trains dep.	a.m.	a.m.	a.m.	a.m.	a.m.	p.m.	p.m.	p.m.	p.m.	p.m.	p.m.	p.m.	p.m.	p.m.	p.m.	p.m.
Granton ... „	7 35	8 35	9* 5	10 20	11 20	12 20	12*44	1 20	2 20	3 20	4 20	5*14	6 40	7 20	8 20	9 3
Trinity... „	7 38	8 38	9 10	10 23	11 23	12 23	12 49	1 23	2 23	3 23	4 23	5 19	6 43	7 23	8 23	9 8
Scotland St. „	8 5	9 5	9 37	10 35	11 35	12 35	1 5	1 35	2 35	3 35	4 35	5 35	7 5	7 35	8 35	9 20
Bonnington... „	8 9	9 9	9 42	10 39	11 39	12 39	1 9	1 39	2 39	3 39	4 39	5 39	7 9	7 39	8 39	9 24
Leith ... arr.	8 12	9 12	9 46	10 42	11 42	12 42	1 12	1 42	2 42	3 42	4 42	5 42	7 12	7 42	8 42	9 27

Trains marked ♥ depart (sooner or later) on the arrival of the Northern Trains.

EDINBURGH TO LEITH.

Trains dep.	a.m.	a.m.	a.m.	a.m.	a.m.	a.m.	a.m.	noon.	p.m.	p.m.	p.m.	p.m.	p.m.	p.m.	p.m.
Edinburgh ... „	8 0	9 0	9 32	10 0	10 30	11 0	11 30	12 0	12 30	1 0	1 30	2 0	2 30	3 0	3 30
Scotland St. „	8 5	9 5	9 37	10 5	10 35	11 5	11 35	12 5	12 35	1 5	1 35	2 5	2 38	3 5	3 35
Bonnington... „	8 9	9 9	9 42	10 9	10 39	11 9	11 39	12 9	12 39	1 9	1 39	2 9	2 42	3 9	3 39
Leith ... arr.	8 12	9 12	9 46	10 12	10 42	11 12	11 42	12 12	12 42	1 12	1 42	2 12	2 45	3 12	3 42

Trains dep.	p.m.	p.m.	p.m.	p.m.	p.m.	p.m.	p.m.	p.m.	p.m.	p.m.	p.m.
Edinburgh ... „	4 0	4 30	4 55	5 30	6 5	6 30	7 0	7 30	8 0	8 30	9 15
Scotland St. „	4 5	4 35	5 0	5 35	6 10	6 35	7 5	7 35	8 5	8 35	9 20
Bonnington... „	4 9	4 39	5 4	5 39	6 14	6 39	7 9	7 39	8 9	8 39	9 24
Leith ... arr.	4 12	4 42	5 7	5 42	6 17	6 42	7 12	7 42	8 12	8 42	9 27

LEITH TO EDINBURGH.

Trains dep.	a.m.	a.m.	a.m.	a.m.	a.m.	a.m.	a.m.	a.m.	p.m.	p.m.	p.m.	p.m.	p.m.	p.m.	p.m.
Leith ... „	6 5	7 35	8 35	9 28	9 50	10 20	10 50	11 20	11 50	12 20	12 50	1 20	1 50	2 20	2 50
Bonnington... „	6 8	7 38	8 38	9 31	9 53	10 23	10 53	11 23	11 53	12 23	12 53	1 23	1 53	2 23	2 53
Scotland St. arr.	6 13	7 42	8 42	9 36	9 57	10 27	10 57	11 27	11 57	12 27	12 57	1 27	1 57	2 27	2 58
Edinburgh ... „		7 50	8 50	9 40	10 2	10 35	11 2	11 35	12 2	12 35	1 2	1 35	2 2	2 35	3 2

Trains dep.	p.m.	p.m.	p.m.	p.m.	p.m.	p.m.	p.m.	p.m.	p.m.	p.m.	p.m.	p.m.	p.m.
Leith ... „	3 20	3 50	4 20	4 50	5 20	6 0	6 20	6 50	7 20	7 50	8 23	8 50	9 35
Bonnington... „	3 23	3 53	4 23	4 52	5 23	6 3	6 23	6 53	7 23	7 53	8 23	8 53	9 38
Scotland St. arr.	3 27	3 57	4 27	4 57	5 27	6 7	6 27	6 57	7 27	7 57	8 27	8 57	9 42
Edinburgh ... „	3 35	4 2	4 35	5 2	5 35	6 12	6 32	7 5	7 35	8 2	8 35	9 2	9 50

Special Trains between Granton and Edinburgh to suit the North Steam-Boats arriving at and departing from the Pier; but to Granton not earlier than 6-15 a.m. nor later than 10-0 p.m.; and from Granton not earlier than 7-0 a.m. nor later than 10-0 p.m. Only First and Second Class Tickets will be issued by the Trains that may run earlier in the morning or later at night than the hours mentioned in the above Table.

WAVERLEY STATION AND SIR WALTER SCOTT

Sir Walter Scott[15], seeking a title with no unwanted connotations for his 1814 novel, chose Waverley (in Surrey, today a local government district with borough status). By virtue of the book's success, Waverley came to denote his entire series of historical romances and he became the first English-language author to achieve a truly international readership in his own lifetime. Edinburgh duly acquired a Scott Monument, erected in 1840–4. The interior spiral staircase reaches to a commanding height with excellent views around Princes Street. Edinburgh also acquired a Waverley Bridge which bestrode the western end of the so-called General Station, where the North British Railway and the Edinburgh & Glasgow Railway came together, with the Edinburgh, Perth & Dundee Railway as a near neighbour.

Waverley Bridge Station (the style which North British and Edinburgh & Glasgow preferred) operated on a joint basis, formally or de facto, as long as the latter company kept their independence. Though Canal Street (Edinburgh, Perth & Dundee) stood apart as the bits-and-pieces layout dictated[16], timetables, notices and advertisements show that 'Waverley Bridge' was understood to embrace it too. 'Waverley' became official after the amalgamations of the 1860s. In similar fashion, Waterloo *Bridge* Station, London, would become in time simply Waterloo; but Waverley's evolution makes an especially complicated story. At first the North British had styled their own terminus 'North Bridge' and from time to time this usage reappeared. 'Princes Street' also came and went – and not only as the alternative to Canal Street – despite the Caledonian's claiming this title for *their* terminus. Donald Cattanach[17] notes how, into at least the 1860s, the Press employed numerous variations, including 'General Terminus', 'Great Central Station' and 'Waverley Termini' (note this *plural*), but 'General Station' prevailed as the accepted, unofficial name. In any case, very little of the earlier fabric was reused in the great reconstruction of the 1890s – by which date there could be no doubt that 'Waverley' would be perpetuated. The last remains of Canal Street were obliterated and today only the building at the south end

---

15 He was created a baronet in 1820.

16 A spur was added, in the Haymarket direction.

17 See Further Reading.

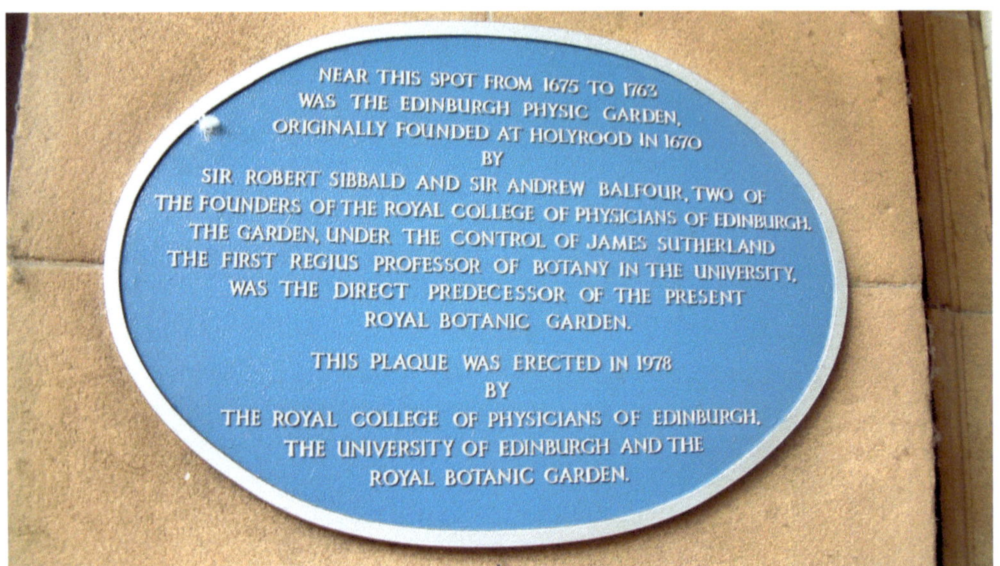

A good deal of Edinburgh's history was buried or otherwise obliterated at the end of the 19th century, when Waverley was comprehensively rebuilt, in conjunction with the new North Bridge.
Robin Howie

of the Waverley Bridge, which serves Lothian Buses as an enquiry office, can claim any antiquity[18].

The North British Company came to realise Waverley's very positive resonance for advertising and publicity. Identifying their route to Carlisle with Scott's novels was a stroke of genius.

Indeed, the popular railway author John Thomas, with pardonable hyperbole, has described Scott as the company's patron saint. For good measure, two successive North British express engines were christened 'Waverley', as was one of their Clyde Coast paddle vessels, in old age (1940) a casualty at Dunkirk. (Its replacement, built for the LNER just before Nationalisation, endures in preservation, as the world's last seagoing paddle steamer – the embodiment of Scotland's maritime heritage. 2022 marked the 75th anniversary of her maiden voyage.) The Trossachs, too, became 'Scott country', accessed from North British Aberfoyle (but also from Caledonian Callander). Though today both these railheads have gone, the veteran steamer 'Sir Walter Scott' still plies Loch Katrine.

Scott can be dismissed nowadays as famous for being famous but otherwise out of date. Yet A. N. Wilson, in a well-received revisionist biography[19], closes his enthusiastic reassessment with this uncompromising tribute: 'We encounter in Scott's novels the greatest diversity of realistic human characters outside Shakespeare; we discover from reading his

---

18 Weatherstone's Waverley Bridge pub, 'The Booking Office' in fact began as a parcels' office in 1898.
19 See Further Reading.

Waverley today obliquely acknowledges Sir Walter Scott's contribution to Scotland's railway history. *Robin Howie*

biography one of the most genial men who ever lived.' Neither British Railways nor their ScotRail successor have done full justice to Scott's memory. During the 1960s and 1970s the former obsessively pursued modernisation. With ex-LMS, ex-Caledonian Princes Street closed and demolished, Waverley should be known, unsentimentally and efficiently(?) as 'Edinburgh', and travellers need learn nothing of the city's railway history! However, after many objections, the time-honoured name was officially reinstated. ScotRail, for their part, have proved ambivalent. The Travel Centre in Waverley's central hall-cum-waiting area offers a bald 'Welcome to Edinburgh'; 'Waverley' has disappeared from timetables; and the electronic displays at intermediate stations on the various routes into the city specify 'Edinburgh'. A prominent plaque commemorates the celebrated LNER locomotive engineer Sir Nigel Gresley, but Scott goes unhonoured, save for an oblique tribute – as it were in instalments – along the elevated walkway linking Market Street and the Waverley Steps. (One of these too-enigmatic notices reads 'Great Scott! Waverley is the only station in the world named after a novel! Thank you, Sir Walter Scott.') Surely less than he deserves? And strictly speaking the name came *indirectly* via Waverley Bridge.

As ever, any citizen needing a taxi to the station simply directs the driver to 'Waverley' – no fuller explanation required!

The various station entrances – from Princes Street by the Waverley Steps (now enclosed and complemented by escalators); from Waverley Bridge, almost within the shadow of the Scott Monument; from Market Street and from Calton Road – are prominently signed

The tower of the North British Hotel (now the Balmoral), framed by the main arches (west and east) of the Scott Monument. *Robin Howie*

'Waverley' or 'Edinburgh Waverley'. The Calton Road approach, whence a pedestrian bridge leads to the east-end platforms, is the least used (and thus the least known) but arguably it has the biggest impact, as one passes under high, wide and handsome Waterloo Place Bridge. Waverley of the early 20th century remained Britain's largest station (disregarding unique Clapham Junction), until overtaken by rebuilt Waterloo in 1922. It was undoubtedly the North British Company's showpiece. Though the four Castle and Bear mosaics in the then booking hall are long gone, a discerning traveller with time to spare can still discover two tall pilasters, inscribed respectively 'NBR' and '1897'.

Nevertheless, the North British were not altogether masters in their own house. The North Eastern Company worked almost all the prestigious East Coast Anglo-Scottish trains in and out of the Scottish capital[20] – which became a particular sore point during the Railway Races to Edinburgh in 1888. When the Caledonian Company brought the West Coast contender triumphantly into Princes Street, the North British could claim no corresponding kudos. With the Forth Bridge completed, they would participate eagerly in the subsequent Race to Aberdeen (1895), during which some startling – and surely foolhardy? – times were recorded between Edinburgh and Dundee.

---

20 The reciprocal bargain made in 1862 gave the Scottish company access to Newcastle via Riccarton Junction and Hexham. It was a dubious quid pro quo, but at that date the North British were intent on consolidating their position all along the Border and into Northumberland.

It remains debatable whether Waverley is in some measure spoiled by the low-set overall roof which had become the only option after the city council refused to relax the servitudes governing the whole site. Perhaps by way of compensation, the North British Hotel (now the Balmoral) was made especially imposing. It stands uncompromisingly above the sprawling station, and there are those who feel that the result is over-heavy.

By contrast, the proportions of Caledonian Princes Street, thoroughly reconstructed by 1894 but swept away in 1969-70, were best appreciated at the station throat – from 1903 the new red-sandstone Caledonian Hotel, ranging along Lothian Road, concealed all 'railway business'. Hotel and station shared a narrow-frontage, and the street-level entry to the latter opened on an angled concourse. The hotel, little altered externally, has retained its given name. Internally Princes Street, with its dark woodwork and distinctive clock, possessed to the very end a pleasing coherence – it was *friendly*, where Waverley's sheer size though always impressive is sometimes intimidating. Nonetheless Waverley today, with steam long banished (excepting occasional heritage 'specials') and diesel traction much reduced, has successfully thrown off the weight of years. Comprehensively electrified tracks lie beneath a clean, refurbished roof and essential signage is up-to-date and uncluttered.

Waverley's western concourse in 2022. Note the interplay of light and shadow thanks to the re-glazed station roof — a very welcome improvement. *Robin Howie*

WALK NOTES

Maps

Even city residents may find themselves unaware of their position relative to familiar streets when they set out to explore Edinburgh's railway past – it can be initially confusing! Residents and non-residents alike ought to take a city map. Only a few sites of the long-gone passenger stations are identifiable by name, only a few by railway remains (bridges, steps, unassuming frontages, decaying platforms…). In many cases no ready evidence at all exists. However, adventuresome walkers will simply start from Waverley Station and follow the walking routes, whether in the prescribed order (which is recommended) or just as they choose. All the routes have a multitude of signposts.

For the ultra-cautious, the 2016 Spokes Edinburgh Cycle Map, 10th edition, 4 inches per mile, clearly shows the former railway lines. A city map of the same scale also serves, albeit the old trackbeds are depicted as cycle routes and walkways.

Although Ordnance Survey Landranger Map 66 covers the city, its 1:50,000 scale is too condensed to trace the dismantled railway network, while for walkers the 1:25,000 Explorer Map 350 lacks the clarity of the Spokes Map. Outside the city, OS Landranger Map 65 is sufficiently clear to cover the walk to Queensferry, the walk to Bangour and the western extension of the Balerno walk. Along with Map 66, it supports the background history of railways in and around Edinburgh which this book includes. For the Bangour Walk, see also Ordnance Survey Explorer Map 349.

David Ross's *The Caledonian, Scotland's Imperial Railway: A History* has a reduced-to-essentials map of 'The Caledonian in Edinburgh', quite unequalled for clarity and with helpful dates. Of necessity, this map includes the Caledonian's North British enemy, so as to illustrate and explain their rivalries across the northern half of the city.

Public Access

The **Scottish Outdoor Access Code** is based on three key principles – **respect for the interests of other people, care for the environment and responsibility for one's own actions.**

With access rights come responsibilities. The Code may be especially relevant on the western extension of the Balerno route (Walk 10). Trackbed walkers, with or without dogs, should take particular care in traversing pasture and remain alert for livestock.

WALK 1

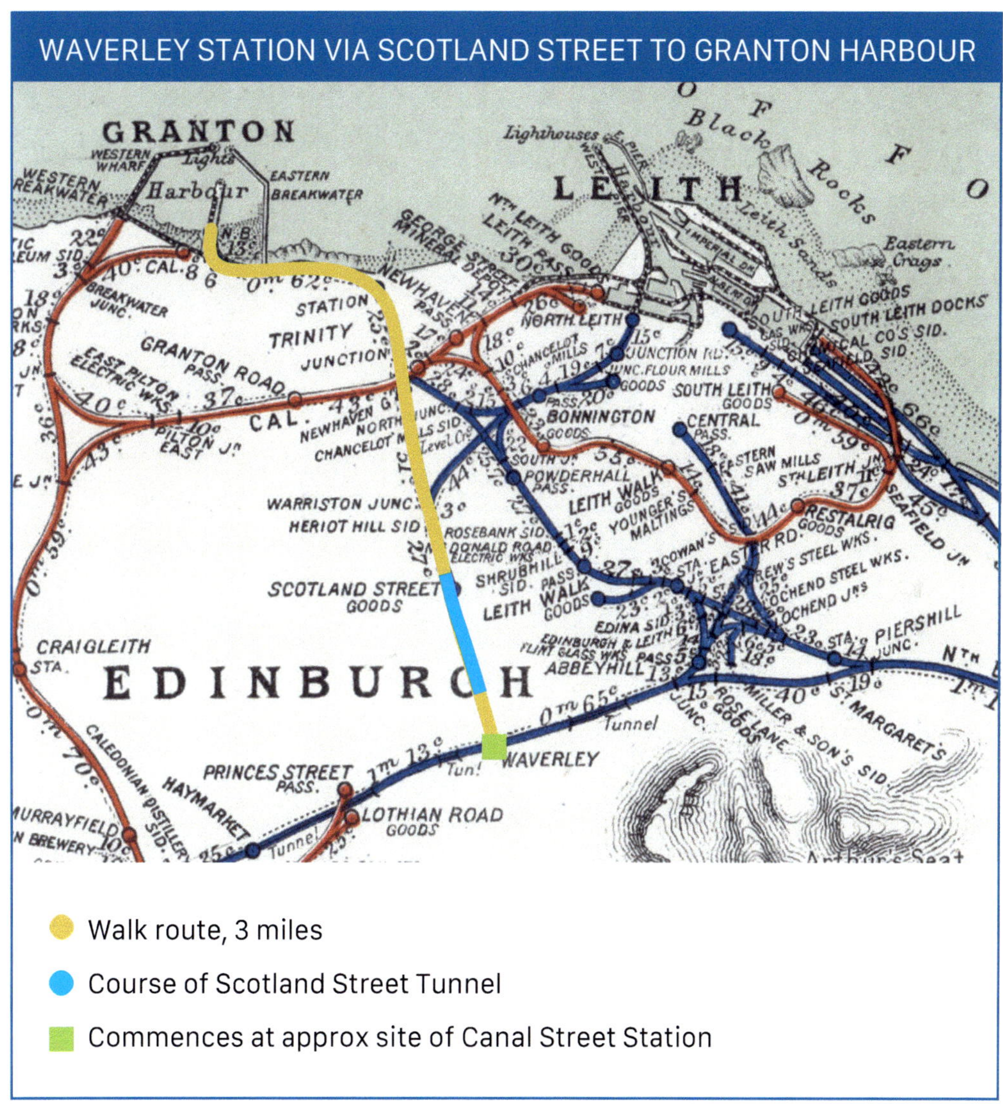

Walk route, 3 miles

Course of Scotland Street Tunnel

Commences at approx site of Canal Street Station

Clutching his ticket for the Hogwarts Express, Harry Potter stood, vainly looking for platform 9¾. 'Just walk straight through at the barrier between platforms nine and ten … best to do it at a bit of a run if you're nervous' was the advice – *Harry Potter and the Philosopher's Stone*, J. K. Rowling.

On the wall backing Waverley's northernmost platform (today's number 20, a relatively recent addition to the layout established in the 1890s) a notice marks the approximate site of Canal Street station – from 1849 the city terminus of the Edinburgh, Perth & Dundee Railway which was absorbed by the North British Company in 1862. (There is a remoter history – the Edinburgh, Leith & Newhaven Railway, subsequently the Edinburgh, Leith & Granton, dated back to 1836. The Edinburgh & Northern Company eventually acquired it as their south-of-Forth connection, and thereupon adopted the title 'Edinburgh, Perth & Dundee'.)

The notice, located just east of Waverley Bridge and behind a high meshed fence, also marks the location of 'lost' Scotland Street tunnel. It is set within a line of blank arches, pierced only by a small ventilation pipe. Sadly, for even a modern-day wizard, it is not possible to pass through to reach any platform, never mind 9¾. The tunnel's southern end was obliterated in 1983 to make way for the foundations of a shopping centre, occupying the site of the earlier Waverley Market, now rebranded as the Waverley Mall.

The location of sometime Canal Street station can still be approximately identified. It stood at right angles to the east-west 'General Station'. *Robin Howie*

From the very beginning, Scotland Street Tunnel severely constrained railway operations. 1,000 yards long, 24 ft wide and 24 ft in height, it dropped north from Canal Street on a 1-in-27 gradient. The roof is 37 ft deep under Princes Street, 49 ft deep under St. Andrew Street and just below street level at its northern extremity. The formidable northbound descent was by gravity, controlled by two brake wagons, whereas, on the southbound return, working by stationary winding engine was inevitable.

Pending completion of the daunting tunnel, the Edinburgh, Leith & Newhaven began operations from Scotland Street station using horse traction. Even after locomotives had been introduced beyond Scotland Street, horse shunting continued at cramped, isolated Canal Street. (A spur was eventually inserted, connecting what was by then the Edinburgh, Perth & Dundee terminus with the Edinburgh & Glasgow Company's extension eastwards from Haymarket.) Despite the expense and effort which the tunnel entailed, it was abandoned in 1868, after only 21

Plaque, north portal, Scotland Street tunnel. *Robin Howie*

Scotland Street station and good yard — the latter long survived the former — lay between the Scotland Street and Rodney Street tunnels. *Robin Howie*

years of operation, when the North British opened a longer but untroublesome line via Abbeyhill to join the original route. This diverged from the Edinburgh–Berwick main line, after the latter's descent through Calton Tunnel.

For today's walker, it is a pleasant ¾ mile surface stroll from Waverley. Ascend to Princes Street. Adjacent to Waverley is the imposing hotel which senior citizens still refer to, with affection and nostalgia, as 'the North British' or 'the N.B'. It was completed by that company in 1902 to complement their redesigned and much enlarged Waverley Station. The latter-day change of name to the Balmoral Hotel raised a few eyebrows. However, one tradition that has survived is having the hotel's prominent clock set three minutes fast – so that travellers would not miss their trains – and the clock only chimes on time on Hogmanay.

Cross Princes Street to the east side of St Andrew Square. After crossing Queen Street/York Place, descend north on Dublin Street and follow the curve of Drummond Place to reach the far end of Scotland Street, where a view over the King George V Park suddenly opens out. Steps lead down and turning right reveals the majestic tunnel portal. What a pity a basketball post is situated right in front!

Robert Louis Stevenson wrote about the arrival of the train into Scotland Street Station:

> The sight of the train shooting out of [the tunnel's] dark maw with the two guards upon the brake, the thought of its length and the many ponderous edifices and thoroughfares above, were certainly things of paramount impressiveness to a young mind.

From 1868 the tunnel was used intermittently for a variety of purposes – it was the city's largest air raid shelter during the Second World War with fresh water, drainage, electricity and space for 3,000 people; moreover, it was used as a radiation testing area by the University of Edinburgh in 1948. Another activity, however, has lingered in the public mind: those asked about the tunnel usually respond – *Is that where mushrooms were grown?* There were indeed several such attempts, ending in 1929.

Scotland Street goods yard survived the passenger station by almost 100 years. The northern end of the tunnel continued to provide coal storage until the 1960s when the tracks were finally lifted. Road vehicles were stored there in the 1980s. For more information, visit the website of Subterranea Britannica, a UK-based society, founded in 1974, for all those interested in man-made underground structures and spaces and how they are employed.

The surrounding area, once idyllically rural, included Canonmills Loch. In hard winters the loch, only 4 ft deep, became the home of the Canonmills Curling Club, claimed to be the oldest of its kind. The large area of water gradually receded, leaving a marshy area known as the Haugh. By 1860 both haugh and loch had been drained, making way for the Royal Patent

Gymnasium (1865) where people could exercise using some ingenious apparatus – which included a giant see-saw for 200 people. The Gymnasium was eventually replaced by the St Bernard's Football Club. The last match on their ground was played in 1942 and the stands demolished in 1947. Difficult to imagine now!

The route to Granton continues by the 180-yards-long Rodney Street (Heriothill) tunnel, the second longest that can be walked in Edinburgh. (At 572 yards, the St Leonard's tunnel takes first place – see Walk 11: Edinburgh & Dalkeith Railway.)

The handsome Rodney Street bore, broad, dry and gently curving, reopened as a walkway in 2009 … pity about the graffiti. Its floor has been raised perceptibly, reducing the original height. Expect a few echoes! In the days of goods traffic and associated shunting, in and out of Scotland Street yard, rumble and tremor was all too apparent to those sitting in the nearby Ritz Cinema (1929) … and to the projectionist! The cinema outlived the railway, closing in 1981.

Beyond the tunnel, the walk continues on the Goldenacre and Trinity Paths. Pass by the Tesco superstore and car park on the right and the brick-built flats on the left. Cross the Water of Leith by the Warriston Viaduct (a splendid triple-arched masonry bridge, best appreciated from water level) to reach Warriston Junction.

The extension to Granton, with change of company title, was approved in 1844, along with the always-intended Leith branch (Warriston-Bonnington-North Leith). The latter, opened in 1846, soon sustained a more intensive service than Granton required (See Walk 5, Warriston Junction to Leith Citadel Station).

Plaque, Rodney Street tunnel. *Robin Howie*

Continue north on what was nevertheless the 'main line', through a deep, secluded cutting. A straight approach, under Warriston Gardens and Ferry Road, leads to what was Trinity Junction, but is now Five Ways. Here in 1868 the line from Abbeyhill via Easter Road and Powderhall came in, on the now designated Chancelot Path (See Walk 2: Waverley via Abbeyhill to Trinity Junction). The Caledonian line to Leith (1862) which today runs eastwards as the Hawthornvale Path had already bridged the North British tracks (See Walk 4: Crew Junction to Leith North). Now that the Caledonian bridge has been removed and the levels adjusted, the result is a general crossroads – and 'Five Ways' makes an appropriate title.

Go straight over as signed (Wardie Bay and Lower Granton ¾ mile and Granton 1¼ miles). A slight descent in a deep, wooded cutting leads to Trinity Tunnel below East Trinity Road; its portals are similar to that at Scotland Street. Continue under the very high Lennox Row Bridge to reach Trinity station, closed in 1925. The platform on the right is the more prominent and behind it is the station building, beautifully preserved as a private dwelling.

The line then curved left and crossed above the double-bend linking Lower Granton Road and Trinity Crescent – despite the traffic lights, today's pedestrians must exercise care to reach the seafront. Only a small part of one bridge abutment remains, and the ongoing embankment has been removed.

The dismantled line to Scotland Street, looking back from Trinity Junction signal box. The Caledonian overbridge is still in situ. *N.D. Mundy, (Stevenson Collection), courtesy NBRSG*

Just east by Trinity Crescent is the eponymous pub-restaurant which marks the site of the Chain Pier (1821). Before 1820, ships and boats used Newhaven Harbour or the two Old Docks at Leith but, with increasing use of steamships and ever-more congestion, another point of call was needed. The Pier consisted of a timber deck suspended by chains mounted on timber towers. Until Granton Harbour opened in 1838 it handled most steamer services, which continued, though reduced, until the 1850s (there had been plans to expand Newhaven's wharfage, in conjunction with the proposed railway, but Granton was developed instead). The Chain Pier was destroyed in a storm in 1898 but the remains of a few of the timber piles can still be seen when the tide is very low.

Only ¾ mile to go! The railway headed west on an embankment between Lower Granton Road and the shore. Follow the roadside pavement, the McKelvie Parade, to reach the broad tarmac way which has covered the trackbed. Note on the left a serpentine row of two-storey dwellings, some colourfully painted and known at one time as East Cottages. Constructed in the 1830s, to house some of the workers building Granton Harbour, these Category-C listed buildings of red Granton brick have surely appreciated in value following the removal of the railway earthworks.

At Granton Square turn right along the Middle Pier to the Royal Forth Yacht Club-house – the site of the single-platform station which closed in 1925. Nothing remains

The former Trinity station, looking west. *Late C.J.B. Sanderson (Stevenson collection), courtesy NBRSG*

The Chain Pier, Newhaven (1821-98) *Tom Wright*

save old railway lines, still to be seen above the stone slipway.

Passenger crossings to Burntisland began in 1844 and from 1846 these were given over to railway-owned paddle steamers, sailing eight times a day in less than 30 minutes. This, however, was inadequate for goods traffic. The problem was solved in 1850 by the experimental *Leviathan*, designed by Thomas Bouch, – a 399-ton paddle steamer capable of carrying 34 wagons on a through deck equipped with railway track. Independent engines kept the deck clear. The Forth's 20-feet tidal difference was ingeniously overcome by special slipways, equipped with cable-operated cradles. The Leviathan, soon with her counterpart on the Firth of Tay, served from 1849 to 1890 assisted by two other steamers. She may fairly be regarded as the world's first train ferry.

A well-loved passenger vessel, carrying up to 950 passengers, was the 'William Muir' (locally known as 'Willie Muir') which entered service in 1879 and finally retired in 1937, having made 80,000 crossings.

Completion of the Forth Rail Bridge in 1890 saw an end to the train ferries, though passenger services continued until 1926. A car ferry continued into the 1960s. The harbour, as first built, was enclosed by two breakwaters. With the western side filled in, it is now bounded the Eastern Breakwater and the Middle Pier and much reduced in size.

The route of the Edinburgh, Leith & Granton Railway can still be traced along the Granton seafront. *Furnevel*

North British Railway paddle steamer *William Muir* at Burntisland – built for the Granton–Burntisland passage, this vessel was withdrawn under London & North Eastern Railway ownership after 58 years' service.

Granton *Square* is unique in that it is the only *street* in the UK so named. The Square is flanked by two impressive buildings in similar style, constructed with stone from Granton Quarry. The west-side building gave entry on the north elevation to the harbourmaster's house and office. On the east side, the substantial Granton Hotel, with stables and a tap room, presumably offered travellers high-class hospitality. During the Second World War, the Navy took over these premises, commissioning them as *HMS Claverhouse*, for reservist training. From 1942 to 1946 Granton harbour, despite its limitations, was home to the shore-based minesweeping training establishment *HMS Lochinvar*. Thereafter decommissioned Royal Navy vessels were scrapped there.

WALK 2

WAVERLEY VIA ABBEYHILL TO TRINITY JUNCTION
GRANTON
LEITH
EDINBURGH
WAVERLEY
Walk route, 2½ miles
Walk commences here

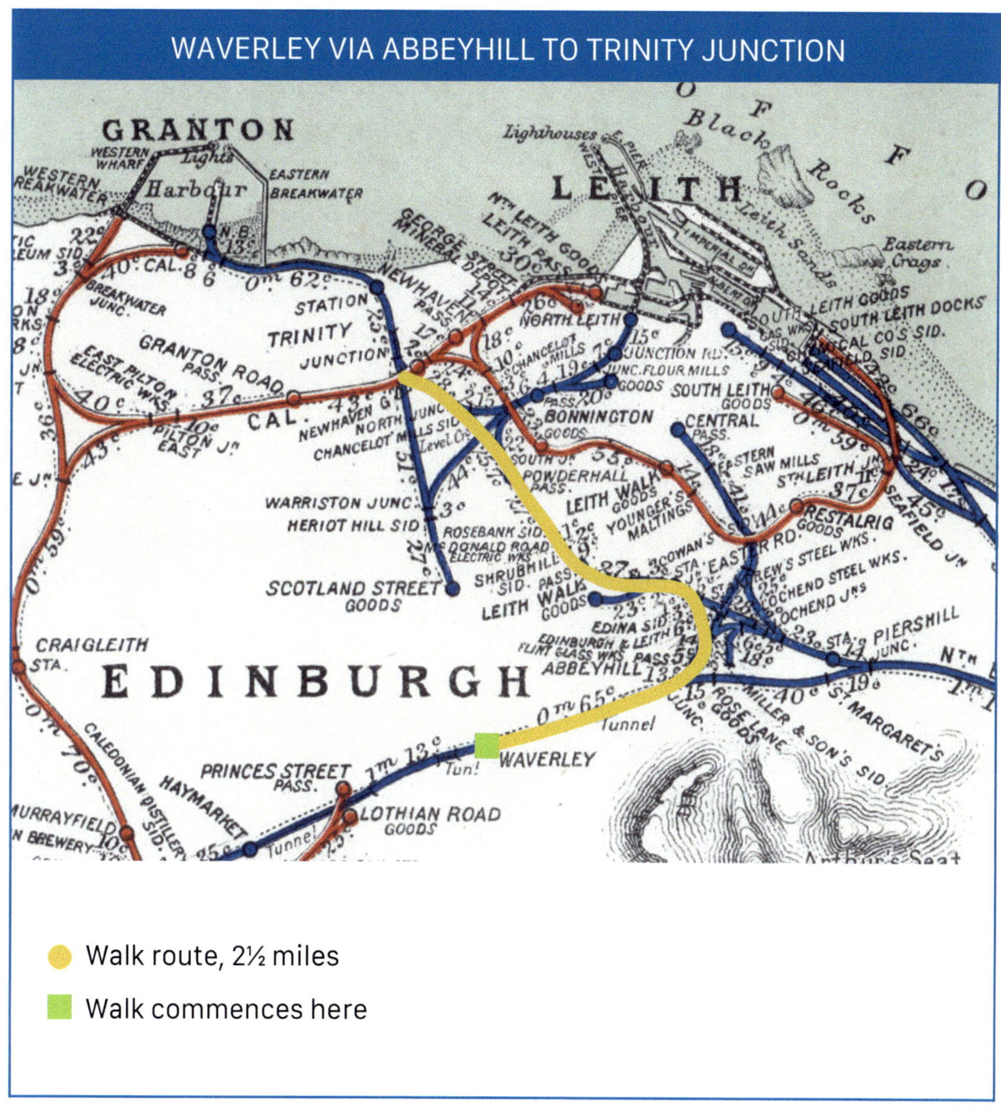

This Walk might be extended to Granton, repeating the final section of Walk 1, so as to recapture something of the 'Ferry Route' as it existed between 1868 and 1890.

From the Waverley Steps leading to Princes Street, head east on Waterloo Place and Regent Road (with grand views of Arthur's Seat) to reach Abbeyhill and join London Road. The railway passed underneath a few yards to the east; note the bridge parapets. The site of Abbeyhill Station, closed in 1964, on the south side by Abbey Lane, is obvious but apart from one platform, covered by vegetation, scant evidence remains. Abbeyhill Junction, further south on Abbey Lane, is clearly identified by the girder bridge. Here, the circuitous but more practical branch to Leith and Granton diverged from the Edinburgh–Berwick main line.

Curving NW to Easter Road station, the trackbed is choked by overgrown foliage, so time for an urban detour. Go down Carlyle Place for an interesting stroll by the Abbeyhill Colonies, one of ten such developments (now conservation areas) constructed by the Edinburgh Co-operative Building Company (ECBC) – model dwellings of the second half of the 19th century to provide improved living accommodation for the working class. The term 'Colony' may derive from the fact that the developments were outside the city when they were first built. They were 'Colonies' in the sense of a community of similar

Abbeyhill station on the new line (1868) to Leith and Granton – from which the later line (1903) to Leith Central diverged. A spur (1888) turned east, for Piershill and Portobello.
G.N. Turnbull (Stevenson collection), courtesy NBRSG

people (artisans). The emblem of the ECBC was a beehive, so perhaps the term derives from workers acting together for the common good.

Pass by Norton Park sheltered housing (the site of the Edinburgh Crystal Glass Co, 1876–1974), then slant NW to Edina Place and the former Edina Works of the map-printing firm, W.& K. Johnston. Now converted into additional sheltered housing, the large two-storey red brick building (1878), with yellow terracotta brackets under the cornice, will draw your attention.

Commemorating 'Edinburgh Crystal'.
Judy Cantley

Reach Easter Road and turn right and later right again on Bothwell Street, at the far end of which is the 1929 Crawford pedestrian bridge. Named after the councillor who proposed

Easter Road and Easter Road station entrance, 1920s. *Tom Wright*

it, the bridge offers a short cut to Hibernian Football Club's Easter Road stadium for away supporters. Straddling both the branch from Abbeyhill and the converging line from Piershill, it offered good views of Easter Road station and Easter Road Junction.

The Piershill-Easter Road Junction line, traces of which remain, crossed Smokey Brae (where Restalrig Road South descends to pass under the Edinburgh-Berwick main line) and ran behind St Margaret's depot, the last redoubt of steam locomotives in 1960s Edinburgh before modern servicing was concentrated at Haymarket and Craigentinny. Meadowbank Sports Stadium came to occupy part of the site.

Reduced to single track, which continued to Powderhall (below) along the abandoned route to Trinity Junction, the Piershill connection saw a considerable traffic in compacted waste to a landfill site in East Lothian. This traffic has ceased, and the redundant railway has now been reserved as a future walking and cycling route in the City Council's Local Development Plan.

From the Crawford bridge, continue by Albion Terrace to the Hibernian football ground. Follow Albion Road round to the left to reach Easter Road. On the east side of the road is a sign, Easter Road Junction, and the very restricted site of the station. Both platforms remain in place though not the street-level booking office. Leith Walk East goods yard was situated on the west side of the bridge. (Apart from the larger goods depots,

A steam 'special' at Easter Road station. Among the array of semaphore signals, two are suspended – as sighting demanded. *NBRSG*

several stations on the city lines had their own little yards, serving local coal merchants and other traders.)

Now proceed as closely as possible to the former railway. Go down Easter Road, turn left on Albert Street to reach Leith Walk and turn left to where the line tunnelled under this important thoroughfare.

A lane on the south side leads to two archways – one blocked, one for the overgrown trackbed. Leith Walk station, on the north side, closed to passengers in 1930 – a casualty of tram and bus competition. Slight traces of the platforms remain, but no buildings.

Edinburgh's modern tramway (Airport to York Place) has now been extended to Newhaven. Construction work on Leith Walk exposed the arched members of what had been in effect twin tunnels accommodating the railway. Even more interesting was the nearby discovery of a pulley wheel dating from 1899, when the city's horse-tram system, first established in 1871, was changed to a cable-tram network with the cable channels sunk between the rails. A series of stationary steam engines powered the endless cables and gave the cars a constant speed of 12 mph; disengaging the 'grabs' brought these vehicles to a halt. (This system was previously adopted on the Glasgow Subway (1896), which had other peculiar features – the gauge was 4 ft and electric lighting was employed from the first, though cable traction endured, as it did in Edinburgh, well into the 20th century.) The discovery of the pulley wheel recalls what was known as the 'Pilrig Muddle' when passengers were forced to change at Pilrig from Edinburgh trams to Leith's then separate system, electrified in 1905. Comprehensive electrification came only in 1922, two years after Leith became part of Edinburgh.

It is then a gentle descent down McDonald Road, parallel to the former line, with two detours to overlook what survives, first when turning right on Dryden Terrace and later from Papermill Wynd – a *street* in all but name, like Granton Square.

Reach the foot of McDonald Road.

Further west is the Powderhall area on the south bank of the Water of Leith. The name derives from a former gunpowder factory and its associated buildings. Powderhall Stadium, built in 1869, was used for many years for greyhound racing and as a motorcycle speedway track. It gained fame for being the place where Olympic athlete Eric Liddell trained in the 1920s. The 1981 film, *Chariots of Fire* – based on the lives of Eric Liddell and Harold Abrahams and culminating at the 1924 Paris Olympics – had an Academy award-winning electronic soundtrack scored by the Greek composer Vangelis. Who could not be impressed by *that* scene, shot at St Andrews beach, as runners splashed through the water … and all the while with *that* music inspiring them that they could, and would, run faster? Played each year circa the London Marathon, it impelled your author to break the 3-hour barrier.[*] The stadium closed in 1995 and the area is now completely occupied by flats. However, the Powderhall Sprint, first held in 1870 – a professional footrace with handicapping – still survives as the New Year Sprint, held at Musselburgh Racecourse.

---

* Robin Howie thrice contested the London Marathon.

Although a path leads straight over the Water to St. Mark's Park (return here later),instead turn right on Broughton Road. Immediately on the left is the entrance to a derelict area. The site was originally a waste incineration plant, constructed in 1893. This, the city's main refuse centre, became in 1985 the site of the Powderhall Waste Transfer Station, where thousands of tonnes of household rubbish were received each year and compacted before being despatched to landfill. The cleared area was used as a construction depot for the tramway extension to Newhaven; it is to become a mixed-use housing development with green spaces.

Head east, passing by a lengthy, red sandstone building. Ornate, Victorian and Category B listed, it is the sole remaining element of the incineration plant and even more impressive when viewed from the sometime construction depot. This building is currently being refurbished as a micro-business hub, the Powderhall Stables, and the former yard at the rear is to be turned into a new plaza to host outdoor events. Adjacent was the frontage of Powderhall station, identified nowadays by the red brick parapets of the overbridge. The station, on the north side, was closed as a war-time economy measure in 1917 and did not reopen afterwards.

Continue further down Broughton Road and enter Redbraes Park, at the foot of which, all of a sudden, is the Water of Leith. A gap in the fence gives a chance to peep through at

The surviving Victorian building which fronted the then Powderhall incinerator. *Robin Howie*

what little remains of the railway bridge over the Water. The double girder span survives, but on the upstream side its cross members, like an exposed rib cage, no longer have any decking. Definitely not a crossing point! I imagine that the entire bridge will soon be no more.

Retrace steps to the end of McDonald Road and follow the tarmac path towards St. Mark's Park. Note on the left the bowling greens. A pedestrian bridge, very high above the Water, leads to the far bank and the southern end of the park. Immediately turn right to join the Water of Leith Walkway, which dips under the railway remains and passes a solitary 'chimney', in fact a sewer ventilator, close by a weir.

Trains bound for North Leith curved right from Bonnington South Junction to Bonnington East Junction. The vanished bridges by which this spur recrossed the Water only to cross back again can be identified by the remains of their masonry parapets, but neither junction is now identifiable (see also Walk 5: Warriston Junction to Leith Citadel Station), albeit the Walkway incorporates a short stretch of the former trackbed, on the north bank.

Just before the weir (above) turn left to a path which passes (left) a forlorn buffer-stop. This marked the end of the reconstituted but once more abandoned Piershill-Powderhall line. An embankment path, overlooking allotments, leads NW by the east side of St. Mark's

Powderhall station entrance in North British guise. *Tom Wright*

The bridge spanning the Water of Leith on the Abbeyhill–Trinity Junction line, seen partially dismantled. The Bonnington South Junction–Bonnington East Junction spur recrossed the Water then crossed a third time to regain the left bank. See also upper image, p.47.
Robin Howe

Park to cross the earlier Scotland Street–Leith line. The tracks intersected at right angles, making a *railway* level crossing ('flat crossing'), where safety was ensured by interlocked signals. To the right is Bonnington station, past Stedfastgate; to the left is Warriston Cemetery. Carry straight on by the signed Chancelot Path in the direction of Trinity and Granton.

From Bonnington North Junction, a spur – curving left to Bonnington East Junction – enabled trains from the Granton direction to reach Leith North without reversal. Housing has occupied the sometime formation such that, as with Bonnington South Junction, identification has become a matter of guesswork.

Beyond the masonry abutments of a demolished overbridge, pass under the Ferry Road and Bangholm View Bridges, both attractively built on the skew. A small curving cutting then leads to the site of Trinity Junction, where the old and new North British routes to Granton converged. You are revisiting the Five Ways interchange! See Walk 1 (Waverley Station via Scotland Street to Granton Harbour).

Powderhall station and Bonnington South Junction. Note Chancelot Mill (right background). The line crossed the Water of Leith just beyond the platform ends. *Stevenson collection, courtesy NBRSG*

The skew overbridge carrying Ferry Road. *Robin Howie*

WALK 3

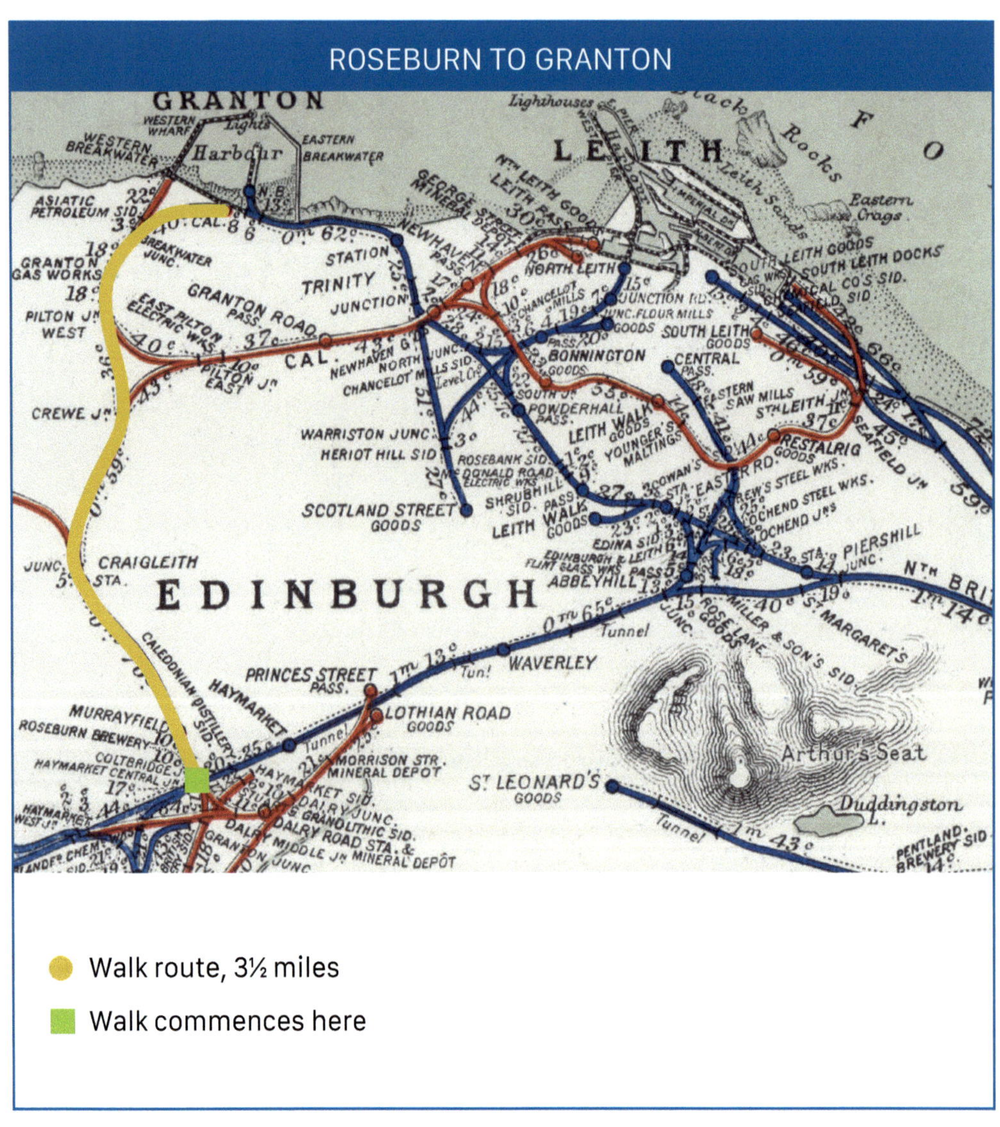

- Walk route, 3½ miles
- Walk commences here

This downhill stroll, via Crewe Junction and Granton Gasworks to Granton Harbour, starts from the corner of Roseburn Terrace and Roseburn Street.

Walk south along Russel Road. Beyond ScotRail's Haymarket Depot (now some 180 years old and several times remodelled), the road takes a sharp turn left. Ahead is a tarmac walkway/cycle path that zigzags to overlook the four-track main line and the Newhaven-to-Airport tramway. The Caledonian Company's viaduct spanning the main line west from Haymarket has been demolished. There was no question, once Princes Street Station had been sentenced to closure, of retaining any local passenger provision on the former Caledonian lines to Granton and Leith. At Slateford and Kingsknowe, Edinburgh-Glasgow trains via Shotts would still call, but under British Railways that route had already been reduced to distinctly secondary status.

Today's walkway joins the broken formation at a high embankment overlooking houses on either side. Note the signpost – 'Granton 3 miles, Leith 5 miles'.

On what is now the Roseburn Path (just under one mile to Craigleith Junction), the difficult terrain demanded two major bridges, high embankments and a deep cutting – altogether a very expensive stretch and including two stations. The rest of this suburban route to Granton seems tame by comparison.

Cross Roseburn Terrace by an ornate girder bridge, more impressive when viewed from the road. As the trackbed curves, pass by lengthy platform walls on both sides … all that is

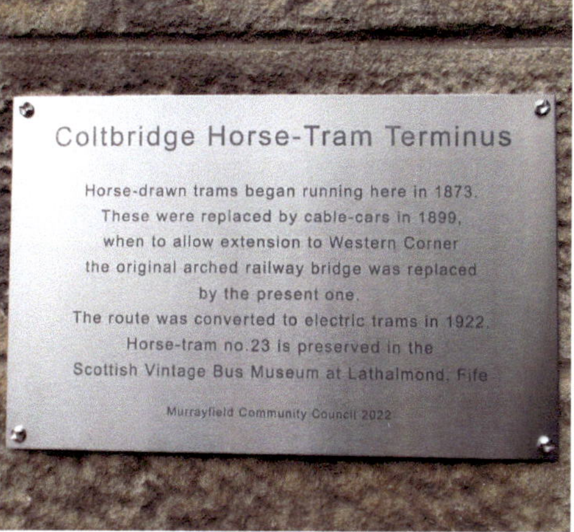

The former Caledonian Railway bridge spanning Roseburn Terrace has a place in the history of Edinburgh's tramways.
Yellowlees

A Leith North–Princes Street train at Murrayfield station. The ex-Caledonian Railway tank engine has a British Railways number and is disfigured by a stove-pipe chimney. *Tom Wright*

left of Murrayfield Station, where 'specials', with dining/buffet cars, were a regular feature on international rugby days.

Beyond, the Murrayfield (or Coltbridge) viaduct crosses Coltbridge Avenue and the deep gorge of the Water of Leith. This three-arched bridge, with masonry piers and abutments, is again best viewed from below. To ensure its long-term integrity, rebuilding works were carried out in 2020 by means of a complex suspended scaffold which supported a working platform beneath each of the arches. I happened to pass by when the initial investigative rope work was underway.

Pass under Coltbridge Terrace and into a tree-lined cutting – possibly the deepest in the city – which takes the former railway under Ravelston Dykes. The cutting gradually comes to an end, to be followed by a high, well-wooded and secluded embankment which leads to Craigleith Station. Only the lengthy curving platforms survive. In 1922 the name was changed to 'Craigleith for Blackhall', reflecting Edinburgh's suburban growth. There were two goods sidings to the east, one of which served Craigleith Quarry.

Pass under Queensferry Road (the trunk A90) and so to Craigleith Junction – the end of the spectacular section – from where a line branched to Davidson's Mains and Barnton. On the right is Craigleith Retail Park, opened in 1993 on the site of the old Craigleith

Murrayfield (Coltbridge) viaduct in present-day condition. *Robin Howie*

Below Ravelston Dykes. *Robin Howie*

The remains of Craigleith station. *Robin Howie*

Quarry, a major source of high-quality building stone used in much of the New Town and in many of the city's landmark buildings. When the quarry closed in 1942 its final massive faces were 360 ft high. The trackbed then traverses much easier tree-lined terrain and passes under the large brick-lined Telford Road Bridge. Later, the tarmac path takes a sharp curve to the left but here the railway continued straight on through a grassy expanse – once a goods yard – at the end of which is a high gated area at one time used by the Fire Brigade. (The Scottish Fire and Rescue Service, Crewe Toll Community Fire Station, is situated at the foot of Telford Road by Crewe Toll.)

A pedestrian footbridge across Ferry Road – replacing the demolished railway bridge, of which only part of the nearside abutments can be identified – leads to the site of the Crewe Junction triangle (Crewe, with Pilton West and Pilton East), which gave connection between Granton and Leith. (See also Walk 4: Crewe Junction to Leith North.) This site has been lost. Only the embankment curving right remains and the area around the pedestrian bridge has been considerably altered by new road layouts.

Continue on the walkway/cycle path – signed 'West Pilton ½ mile' and 'Granton 1½ miles' – which eventually becomes the new West Granton Access road, leading to the

Today's pedestrian bridge over Ferry Road. *Robin Howie*

Waterfront area. It is then a steady descent and into view comes a solitary gasholder (of which more below). Reach a short, well-lit, underpass which takes today's tarmac path under West Granton Road.

However, time for an interesting diversion. Instead of going through the underpass, cross West Granton Road and stroll along Waterfront Broadway. On the right, amidst a fenced-off area, is an impressively large red brick-built building with yellow trim – the former Granton Gasworks Station built by the Caledonian Company, who carried workers free of charge from Princes Street Station. The City Gasworks (later styled Granton Gas Works) opened in 1906, and rail access was vital in bringing coal from across the Lothians and further afield. (Trains arriving from Caledonian-served pits in the west had direct access to Granton.) By 1926 the gasworks were consuming 200,000 tonnes each year.

The station closed in 1942 but still extant are a truncated platform and a clock, stuck at 8.40. Planning permission and listed building consent for the full refurbishment of the station, a Grade B listed structure, has now been given – 'to provide a fully accessible and adaptable building designed to attract and retain modern business and innovation occupiers.' A National Transport History Trust plaque has been placed at ground level.

Granton Gasworks station (1902–42).
Robin Howie

Further along Waterfront Broadway is the Centrica, Scottish Gas Headquarters building, appropriate given the site's history as home to Scotland's largest gasworks. The extensive area which the plant occupied is fast becoming an entirely new Edinburgh district of mixed development – commercial, residential, educational and leisure; as to the last, Forthquarter Park is a new recreation ground.

Cross Waterfront Broadway to the solitary B-listed Gasholder – more accurately, its very large circular steel frame. Built in 1903 and last used in 1987, it stands within an extensive fenced-off area. This large 'skeleton' dominates the scene yet evinces an extraordinary delicacy, in superb contrast to the multitude of new-builds in the area. Thank goodness the campaign to save it proved successful – can modern architecture compete? The other two Granton Gasholders were demolished and removed by 2004. At the time of writing, the historic 157-ft-tall structure, a striking feature of Edinburgh's landscape for more than 120 years, was set to be transformed into a tree-lined amphitheatre for cultural events on the city's waterfront – the centrepiece attraction of a new 'sustainable coastal town'. The new Gasholder Park opened, on schedule, in 2025.

Take a wander through the eight hectares of landscaped open space with boardwalks and water features which have become home to a wide range of wildlife – a remarkable transformation. Native trees and plants have been used and the water features function as a sustainable drainage system. Note the warning sign: 'Danger, Deep Water'. Do make the detour south to the dramatic Granton Pond adjacent to West Granton Road, where you are warned, sensibly, against diving, swimming or fishing.

Return to the West Granton Road underpass, whence the tarmac way, mostly following the line of the railway, heads straight NE to Waterfront Avenue and Saltire Square. Just before that point, the line curved eastwards but is now lost within an industrial area. Instead, follow the broad Waterfront Avenue pavement, signed 'Granton Harbour ½ mile', and pass by new flatted housing to reach what is West Harbour Road. Turn right and pass by a large brick building that housed the shore-based Granton Lighthouse, built in 1874 as part of the Northern Lighthouse Stores and Buoy Yard. Never operational, it was used both for training purposes and to test new lights before these were transported to their

Granton's surviving gasholder frame. *Robin Howie*

Granton Customs House. *Robin Howie*

often-remote working locations. The complex of sheds and stores is currently used for a variety of purposes – for example, Theatre Stuff, a props and costumes hire company, is one occupier.

The Granton Waterfront development is Scotland's biggest brownfield regeneration scheme, and Edinburgh Council say that Granton Lighthouse will be brought back to life, becoming a hub for musicians, artists and creative businesses, with an exhibition centre and café. Funding of more than £2¼ million has been pledged by the Scottish Government.

Just before reaching Granton Square, note the fine masonry building, dating from circa 1860, which served as the Granton Custom House. The name is above the door.

WALK 4

CREWE JUNCTION TO LEITH NORTH STATION
GRANTON
LEITH
EDINBURGH
Walk route, 3 miles
Walk commences here

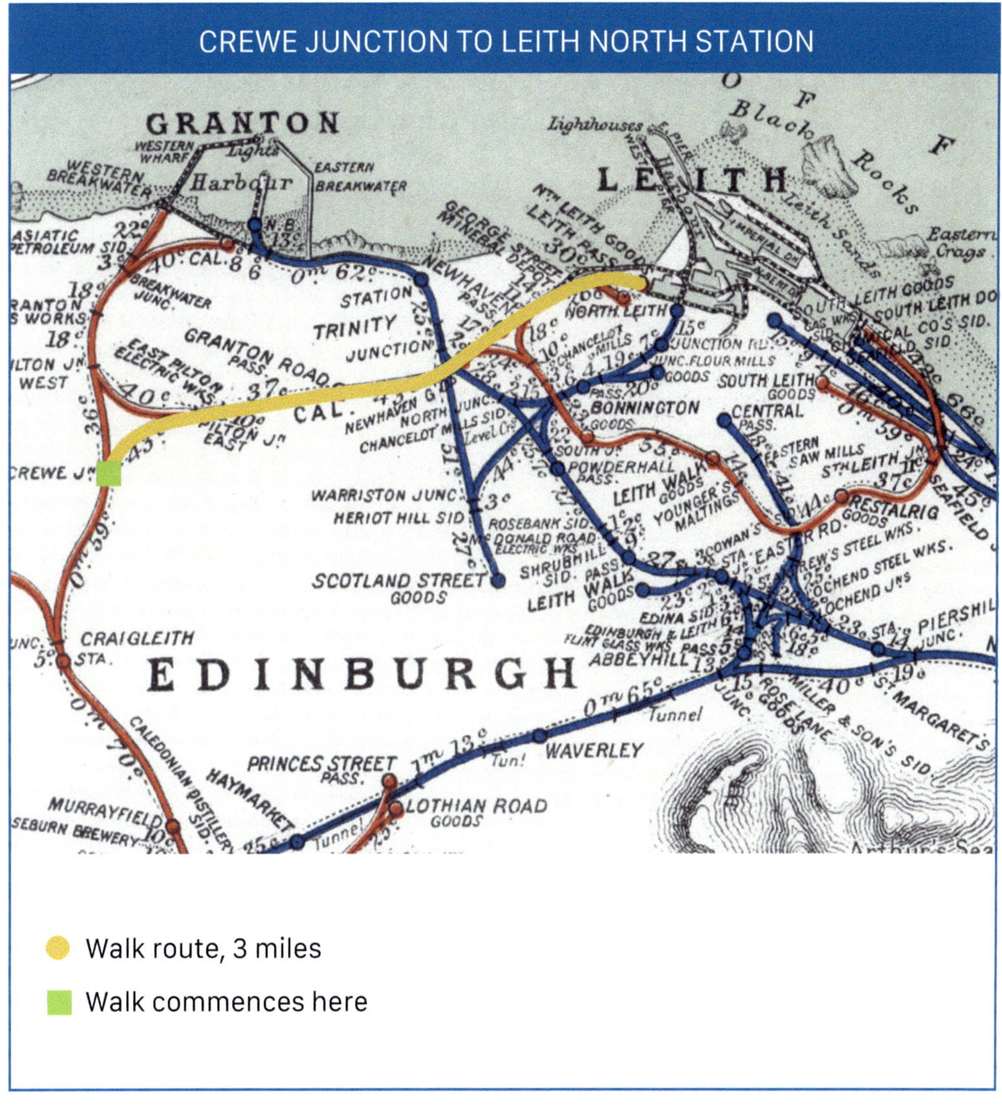

From Crew Junction (see Walk 3: Roseburn to Granton) another sign points east – 'Pilton 1 mile, Trinity 1½ miles, Newhaven 2½ miles, Leith 3 miles'. Follow the curving embankment gently downhill. (Cyclists can free-wheel for more than a mile.)

No evidence remains of the short-lived (1934–62) East Pilton Station, situated between Crewe Road North and the bridge over Pilton Drive. Initially a Halt, it was built to serve the Bruce Peebles engineering works on the left. (To the right is Morrison's supermarket and extensive car park, once the site of the Northern General Hospital.) In the 1950s the works employed more than 3,000 people. However, in 1999 a generator blew up inside the factory and the resulting fire tore through the plant, sending black smoke and ash into the air. Streets were sealed off and people evacuated from their houses. It took weeks before residents were allowed back home. I recall passing by soon afterwards and the smell of burning was pungent. The land at East Pilton has now been redeveloped as housing.

Pass under first Pilton Drive then Boswell Drive, whence there is a lovely wooded approach to Granton Road Bridge, where steps rise to the road. On the eastern side, overlooking the trackbed, are two bricked-up entrances which once led down to Granton Road Station. The station building itself was at road level and at right angles to the line. I have vague memories of the building latterly used as a workshop – my great-aunt and uncle lived on the far side of Granton Road – but all evidence has now gone. Return to the tarmac path which rises slightly to the right-hand platform. (The left-hand platform is almost completely covered in ivy and the trackbed is overgrown.) The path then curves left to follow the trackbed.

Now mostly in a cutting, pass under Wardie Road, South Trinity Road and Clark Road. The arches of the latter bridge, brick in-filled between steel girders, look oddly unattractive. Later, on the left, is a wooden building, used by the 142 (2nd Edinburgh) Squadron Air Training Corps.

The trackbed then descends to Five Ways, today's interchange for walkers and cyclists (again!). That the Caledonian line to Leith bestrode the earlier North British line from Scotland Street to Granton, while the 'improved' North British line came in at almost the same location, complicates any attempt to imagine this railway scene as it once was.

The Caledonian line (1864), originally goods only, was later widened to four tracks through Newhaven and Leith, two to the docks and two to their Leith terminus – passenger services began in 1879. Left of the tarmac way tracing the passenger line, a wide area of silver birch and wandering paths has invaded the sometime goods formation.

Reach Newhaven Junction. Slanting right is the Victoria Path, which follows the Caledonian's late-in-the-day line to Leith East. However, continue straight on as signed, 'Newhave ¾, Victoria Quay and Western Harbour 1¼ miles'. A slight descent leads to the large Craighall Road Bridge; its central masonry pier separated the freight and

Beyond vanished East Pilton station, the route goes straight to still-identifiable Granton Road station. *Robin Howie*

The site of Granton Road station, looking west. Today's walkway-cum-cycleway has obliterated the city-bound platform. *Furnevel*

Straddling the North British tracks just south of Trinity Junction (see also Walk 1: Waverley Station via Scotland Street to Granton Harbour), the line continued to Newhaven station, where the Caledonian station building is still readily recognisable. *Robin Howie*

passenger tracks. Beyond the bridge is a surviving lengthy platform to the right and the remnant of a central platform to the left – marking sometime Newhaven station, where the trackbed area is much dilapidated. Look back to road level to see the wooden station building, which stood at right angles to the line. Two wooden gates mark the spot from which long-gone steps gave access to the platforms.

On paper, between 1902 and 1917 Caledonian Newhaven sat within a triangular junction, but it is questionable whether, earthworks apart, the east-to-south chord giving direct access from their North Leith layout to their Leith East extension was ever laid in.

Craighall Road can be reached from the Victoria Path. The clearly signed Newhaven station house, 85 Craighall Road, has become

Inside old Newhaven station, the booking office 'ticket window' has been preserved. *Robin Howie*

commercial offices. Within the building is a 1961 timetable showing trains to Princes Street (in railway terminology, Up trains), calling at Granton Road, East Pilton, Craigleith, Murrayfield and Dalry Road, together with the corresponding Down trains to Leith North.* Also retained is the booking office 'window'.

The fishwives of Newhaven regularly used the trains, for speedier access to the centre of Edinburgh – just as the fishwives of Fisherrow had begun to do from the 1830s, taking advantage of the Edinburgh & Dalkeith Company's branch. One historian has noted how, into the 1920s, these 'brawny [ladies] with their fragrant creels…might still be seen in a Scottish railway carriage'[21].

Return to the railway and pass under Newhaven Road Bridge, again with a central masonry pier. Further on the left is the well-established Edinburgh Sculpture Workshop, 21 Hawthornvale, an artist-run organisation. The Workshop's modern building, with purpose-built spaces, is open to anyone interested in learning about or making sculpture. Have a rest at the adjacent Milk Bar, watching artists at work, and perhaps contemplate a new interest.

Continue on the Hawthornvale Path to a widening area and the grey-painted North Fort Street Bridge, with two central piers; it once carried the coast road, Lindsay Road, which took

---

* Caledonian Leith was so designated by British Railways from 1952
21 C. Hamilton Ellis, *The North British Railway*, 1955

The wide approach to the Caledonian Company's Leith passenger terminus can still be appreciated today. *Roberton*

On Edinburgh's Princes Street – a tram bound for Newhaven soon after the opening of the extension from York Place via Leith Walk. *Robin Howie*

a dog-leg to the south at this point. However, this broad area, the last evidence of the railway, is **not** the site of Leith North Station.

Pass under the bridge, where a slight rise leads to the realigned Lindsay Road, along which Edinburgh's tramway extension to Newhaven now runs. There are now two Lindsay Roads – the main road and the other a foreshortened curving line of tenements – a layout best appreciated from the North Fort Street Bridge. Passenger and freight lines passed under that bridge and curved east parallel to the north side of the main road. The extensive goods yard has been completely subsumed by the road realignment carried out after 1962 and by the considerable changes to the immediate dockland area in the last 50 years. Just west of the road to the Ocean Terminal, flatted dwellings have occupied the station area.

No doubt local folk readily distinguished between Leith's 'Caley' and 'NB' termini, though strangers must have found the signage 'North Leith' and 'Leith North' (becoming 'Leith North' and 'Leith Citadel' at the very end) more than a little confusing. (See also Walk 5: Warriston Junction to Leith Citadel Station).

The remains of Leith North, where a passenger service – first Caledonian, then LMS, then British Railways – endured into the 1960s. *Furnevel*

WALK 5

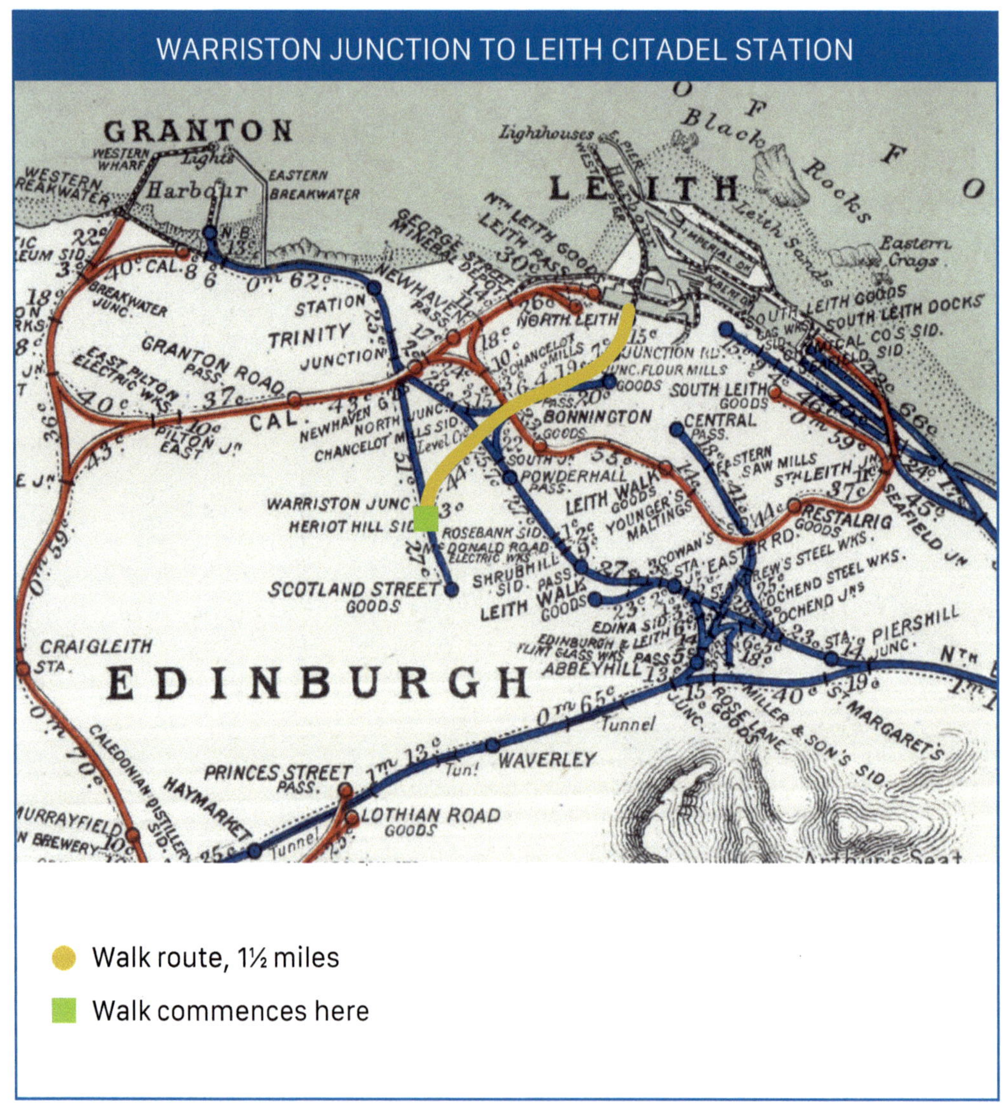

Walk route, 1½ miles

Walk commences here

Cross the Water of Leith by the Warriston Bridge (see Walk 1: Waverley Station via Scotland Street to Granton Harbour) to reach Warriston Junction, where the Leith line slanted right. The route is now signed Warriston Path – St Mark's Park, Bonnington 1 mile, Leith 1 ⅓ miles.'

Immediately ahead is Warriston Cemetery, laid out on land acquired from the Heriot's Hospital Trustees (owners of Warriston estate) by the Edinburgh Cemetery Company (1840). The first interment was in 1843. This elaborate necropolis exemplifies the mid-19th century 'Garden Cemeteries Movement', in which calculation of profit and concern for Public Health came together, at the same time as teeming inner-urban graveyards were being closed. The railway was permitted to cut through the as yet partially developed cemetery – it seems without controversy; an accommodation underpass, built in 1845 and described as 'sober but ornate' (note its Gothic parapets), linked the divided burial grounds. (It was pleasing to see on my most recent visit that the latest graffiti had been erased.)

Consider making a detour to this interesting site. Return to Warriston Junction, head north on the Goldenacre Path and pass under what is the Cemetery access road. Just beyond, steps on the left lead to the end of Warriston Gardens and the Cemetery entrance. Occupying a massive area, it contains tens of thousands of graves. Amongst those interred are notable Victorians and Edwardians – perhaps the most eminent being the physician Sir James Young Simpson – and less notable figures such as my great-grandfather and some of his descendants. Certain areas have become heavily overgrown, but restoration has begun.

Warriston Cemetery Bridge. *Robin Howie*

Beyond the underpass, the silver birch-fringed way passes under Warriston Road and past the entrance to St. Mark's Park. Continue straight on by the extensive Warriston allotments. Cross the route from Abbeyhill, which superseded that from Scotland Street (see Walk 2: Waverley via Abbeyhill to Trinity Junction). Continue beyond vanished Bonnington East Junction towards Bonnington station, passing the spot where the Caledonian Company's line to Leith East (see Walk 6: Newhaven Station to Leith East) crossed both the North British formation and the Water of Leith. The viaduct's two circular masonry piers cannot be missed. To the left, at the same spot is Steadfastgate (described in Walk 6: Newhaven Station to Leith East).

A deep cutting then leads to Bonnington station, easily identified by its lengthy platforms, which have survived. The Newhaven Road overbridge was reconstructed and widened in 1934, and this girder addition clashes unpleasingly with the original arches. Stone steps on both sides lead to Newhaven Road where number 94 was the station house, dwarfed in its later years by four-storey tenements. It has been enterprisingly transformed into the frontage of a modern dwelling (styled The 'Old Train House'), which became Scotland's Home of the Year, 2023.

Return to the Path and look out for a small stone marker dated '1860' below the platform edge. Pass by an unattractive stretch – a short canyon between high walls and buildings. To the right, on the hidden north bank of the Water of Leith, a large area of new-build flats is under construction, alongside old industrial sites by cobbled West Bowling Green Street.

However, after South Fort Street bridge views open out to disclose the widening estuary of the Water of Leith. Pass under the elegant elliptical arch of lengthy Great Junction Street overbridge, which spanned both the railway and the Water. (Another latter-day steel addition blights its western side.) Just beyond and on the left was Junction Road (latterly Junction Bridge) station, of which no evidence now remains. The limitations of the site dictated a single platform, used by trains in both directions – an unusual arrangement on a double-track line.

'1860' marker, Bonnington station. *Rhona Fraser*

Great Junction Street overbridge. *Robin Howie*

At Junction Bridge station (Great Junction Street), on the former Edinburgh, Leith & Granton Railway, a veteran North British locomotive heads a steam 'special'. Curved destination boards were a North British feature, but 'Leith Citadel' (for 'North Leith') is a later coinage. *Hennigan collection, courtesy Lynn*

Great Junction Street overbridge, with the booking office of Junction Road station at centre left. *Tom Wright*

Now a conundrum! The tarmac path slanting right to the Water has definitely nothing to do with the railway, and steps on the left lead to cobbled Coburg Street; an area of old buildings. In fact, the railway headed NE by tunnel. Only 139 yards long, the Coburg Street (or North Leith) bore passed under Coburg Street and Couper Street to reach Citadel Place. Today one can only guess where its portals once stood.

Ascend to Coburg Street, and so to its corner with Great Junction Street. Slant down and over Coburg Street to the twin pillars of the girls' and infants' entrance to the former Couper Street Primary School. (The boys' entrance was from Admiralty Street to the rear.) Opened in 1890, the school closed in 1951. It was demolished, along with most of the other buildings in Couper Street, in the 1970s and the area is now subsumed by the high-rise housing development seen ahead. However, note on the left the School House, with a bell in a recess.

Turn right on Couper Field, then left on Couper Street (noting the interesting melange of high-rise flats and old buildings). From here steps descend to a very high wall on the left – where the tunnel ended. A lane on the right leads to Citadel Place where the final few yards of the old railway can be identified. Reach Commercial Street and turn right, passing by a single-storey masonry building in classical style – North Leith Station. Category B listed, it now serves as the Citadel Youth Centre.

To obviate any confusion with Leith North Station it was renamed 'Leith Citadel' in 1952. Nationalised British Railways made a determined bureaucratic assault on ambiguous

North Leith station building, which, externally, endured in Edinburgh, Leith & Granton guise under North British and then LNER ownership. *Robin Howie*

The 'Eastern Port' of Leith's Cromwellian Fort. *Robin Howie*

or otherwise confusing signage, but rationalisation of what had been competing LMS (Caledonian) and LNER (North British) services made very slow progress and would be overtaken by the closures of the 1950s and 1960s. North Leith became Leith Citadel only in 1952, five years after passenger trains had ceased, while its ex-Caledonian counterpart, where passenger traffic continued until 1962, was renamed Leith North.

The change of name also acknowledged associations with Oliver Cromwell and his Protectorate. The station yard is occupied by Tiso, outdoor clothing and climbing gear specialists, and at the east side of their car park is an old vaulted pend with an arched gateway at either end. After defeating the Scottish army at Dunbar in 1650, Cromwell occupied both Edinburgh and Leith, erecting a new fort or citadel at the latter. (Completed by 1656, this never saw conflict and was turned to industrial purposes after the Restoration of Charles II in 1660.) The Citadel – the name endured – gradually fell into disuse; all that remains above ground is this archway, marking its Eastern Port (Gate), which opened on to what is now Dock Street.

WALK 6

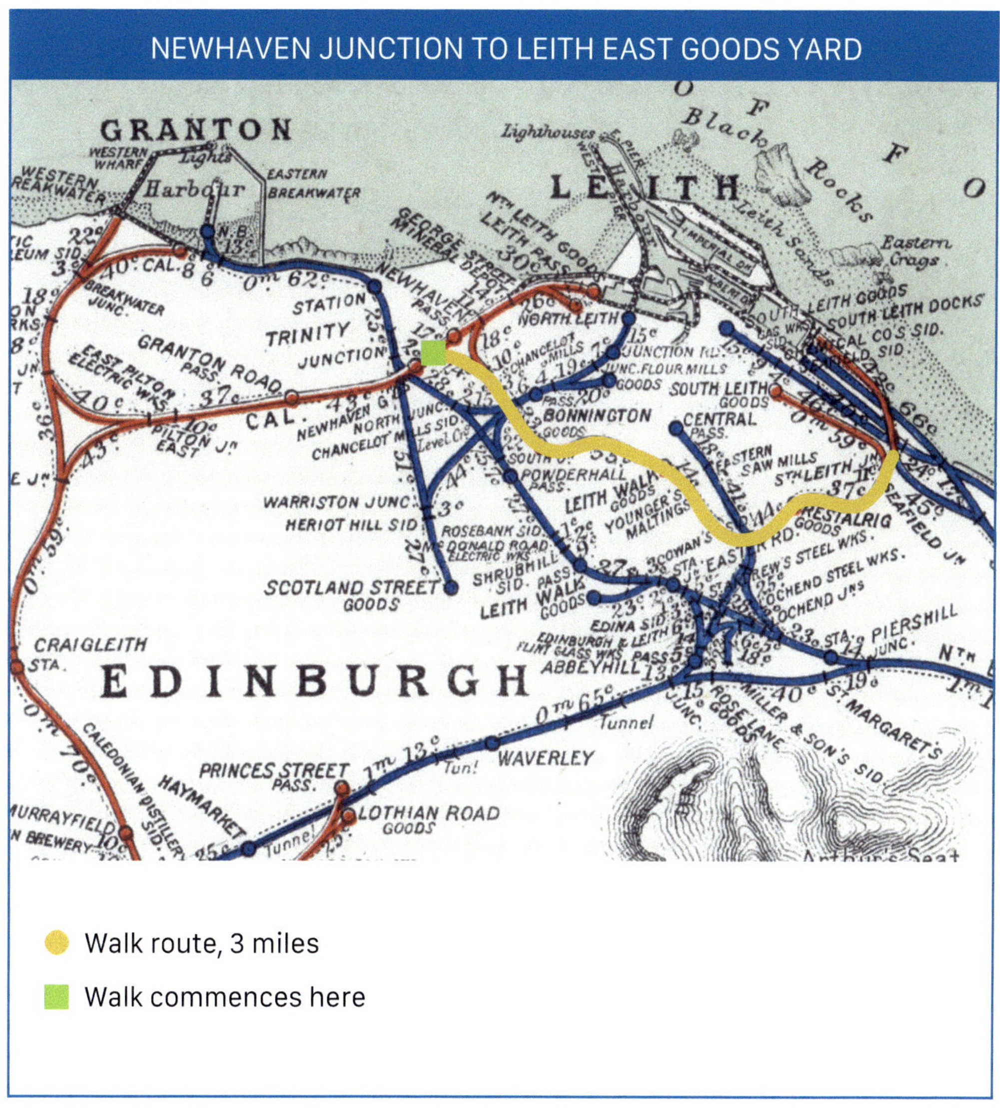

Walk route, 3 miles

Walk commences here

The Caledonian Company's Newhaven-to-Leith East line was intended to have passenger stations at Newhaven, Ferry Road and Leith Walk; these, though almost completed, never opened. Apart from a few 'specials', the line (singled in 1917) handled only general goods and dock traffic. In the OS map of Scotland 1921–30 (One-inch 'Popular' edition), it appears as a mineral line with no stations.

The railway took a wandering route to the eastern quarter of Leith before curving NW to Leith Harbour and Docks – a route accommodating several goods yards and industrial sidings, which demonstrated 'public utility' and so parried some of the North British Company's objections in Parliament.

From Newhaven Junction, where the ground has been slightly raised, follow the Victoria Path and pass under Craighall Road Bridge (a short distance south of the likewise named bridge at the Caledonian's first Newhaven Station) to reach Victoria Park – see the side excursion suggested in Walk 4: Crew Junction to Leith North. The tarmac way rises to enter the park, then descends on exit – a consequence of the 1983 environmental improvement programme, which involved the redesign of the park's footpath network, the planting of trees and the infill of the railway cutting that bisected this public space. In the park is a sign, 'Bonnington 1 mile' and at the exit, on a wall to the left, the sign 'Steadfastgate'. To the south-west is the Chancelot area. (Chancelot Mill

A pier of the Water of Leith viaduct on the Caledonian Company's 'folly' extension to Leith East. From the south bank, the line continued on masonry arches. *Robin Howie*

closed in 1969 and was subsequently demolished. A new mill of the same name has been built at the Western Harbour, Leith Docks.)

Pass under Ferry Road and by Stedfastgate to a spacious grassy way. Just before the tarmac path descends to the Warriston Path and the Water of Leith, veer left to where the Caledonian line ran – here is the Steadfastgate Monument, erected on the Water of Leith Walkway to mark the centenary of the Boys Brigade (1983). At this point the Caledonian viaduct spanned both the much earlier Scotland Street–Leith line and the Water. Its height is indicated by a circular masonry pier on the near bank. (The twin pier a few yards downstream, covered in ivy, is not immediately recognisable.) On the far bank all trace of the railway has gone and the area is now covered by houses.

With no way over, it is time for a short detour, heading left on the Warriston Path to Bonnington station. Use the steps on the right to reach Newhaven Road, which dips to cross the Water by the masonry bridge built 1902–3 and enters the Bonnington area. In 1617 the land and mills at Bonnington became part of Edinburgh when sold by the Logans of Restalrig and for many years the village of Bonny Toon marked the boundary between Edinburgh and Leith.

Immediately on the left is the site of the historic but long demolished Bonnington Paper Mill, replaced by a modern flatted development ('Bonnington Mill'). Behind is a truly massive area of new high-rise flats which have subsumed this old industrial district,

The restored water wheel, Bonnington Mill. *Robin Howie*

leaving just a few surviving tenements such as Ashley Place. Turn right on Bonnyhaugh and continue right, through Milnacre, to the south bank of the Water, to re-encounter the columnar masonry piers from a different view point.

Returning to Bonnyhaugh, slant right tto Bonnyhaugh Lane and so to the Watermill which was moved to this site following its 1980s restoration. The undershot wheel sits on a small section of the old lade, in which the flow of water was once controlled by sluice gates at Redbraes Weir. The wheel is the only surviving example of dozens that once generated power to mill grain, weave cloth, or process leather and paper. A Dutchman, Jeromias van der Heill, was recruited in 1621 to teach his craft of dyeing. The house built for him, named Bonnyhaugh by a later occupant, has been saved, restored and converted into private apartments.

Its route now obliterated by Stewartfield Industrial Estate and more flats, the railway slanted obliquely towards Bonnington Toll, the very busy junction of Newhaven Road and Bonnington Road with Pilrig Street. (Bonnington Road became a toll highway at the end of the 18th century, hence the name.) Continue on Bonnyhaugh Lane, now cobbled, to return to Newhaven Road and so to the Toll, where all trace of the massive railway bridge has gone, save for remnants of an abutment, adjacent to Bonnington Trade Centre. Briefly head south on Pilrig Street and turn left into the pleasingly grassy Pilrig Park. Incidentally, I recall a family outing when a young lad in 1945 to a circus in the Park – the Royal George Circus. Surprisingly, my twin sister has no such recollection!

Excavations in 2006-07 unearthed the remains of Somerset's Battery, one of the two major emplacements linked by the trenches encircling Leith in 1560, when the French garrison were besieged by an Anglo Scottish force. The ensuing Treaty of Leith was signed by Scotland, England and France and presaged the Union of the Crowns in 1603. It has come to be seen as ending Scotland's Auld Alliance with France.

Veer left to charming Pilrig House, built in 1638 and a late example of a traditional Scottish laird's dwelling. It was purchased in 1718 by James Balfour with money received in compensation for his losses made in the failed Darien expedition. Landed investors were reimbursed in full as part of the Parliamentary Union settlement in 1707. As Robert Burns wrote, 'Bought and sold for English gold, such a parcel of rogues in a nation'.

Lewis Balfour, maternal grandfatherr of Robert Louis Stevenson, was born in the house in 1777. It remained the home of the Balfours of Pilrig until 1941, by which time the building had become uninhabitable, although its historical importance led to a Category A listing in 1966. In the mid-1980s the original L-plan tower house was restored, returning the exterior to the original form of 1710. (The name 'Pilrig' may derive from a tower (peel) at the end of a field (rig).)

On the ornate frontage (far side) of Pilrig House is a plaque with a quote from Stevenson's 'Catriona', the sequel to 'Kidnapped': 'I came in view of Pilrig, a pleasant gabled house set by the walk side among brave young woods.'

An early 20th century view of Bonnington Toll, where the girder railway bridge testified to the cost of what soon proved a very dubious project. *Tom Wright*

Pilrig House. *Robin Howie*

The railway embankment followed the northern edge of the park, roughly parallel to Bonnington Road, and then curved eastwards towards Leith Walk. Only traces remain, around the Park's north-east margin. From there, follow a tarmac path past a child's play park on the right and new houses on the left – the latter have obliterated the embankment, but a rise in the path sees it soon resume. Although this leads direct to Leith Walk, it is gated-off and there is no way down at the far end. Return to the path and pass through an industrial area – once the Leith Walk West goods yard – to reach Steads Place. Turn left down Leith Walk to where (at Jane Street) the railway obviously crossed. The high abutments survive but alas the bridge has been removed. (A longer approach to Leith Walk is to exit Pilrig Park heading south-east to Balfour Street, once a tree-lined avenue.)

Cross Leith Walk to Manderston Street, which shortly becomes Gordon Street. The latter is adjacent to the line of the railway –a high viaduct still stands, its arches given over to commercial purposes. The viaduct ends abruptly on reaching Halmyre Street, after which new houses have subsumed the line. Cross Easter Road (the railway bridge has gone) to the start of the Restalrig Railway Path. It is signed 'Lochend Park ½ mile'. The tarmac path

The substantial girder span over Leith Walk. On to Leith East, the line also bridged the North British branch to Leith Central, arguably another 'folly' project. *Roberton*

Today's Restalrig Railway Path, with a dusting of snow. *Robin Howie*

gradually ascends and follows the former line more closely. On the left is Leith Academy and its sports grounds and ahead can be seen Easter Road stadium. From the point where the North British branch to Leith Central was crossed (difficult to imagine nowadays!), the Restalrig Railway Path gently descends some 100 ft to Seafield (see also Walk 7: Abbeyhill Station to Leith Central).

Beyond what may be a wet area in a cutting, pass under the large Lochend Road Bridge (product of the Brandon Bridge Building Co. Motherwell 1900). With rocky outcrops on the right and houses on the left (the latter once the site of Restalrig goods yard), the charming, tree-fringed Restalrig Railway Path reaches another potentially wet area at a second bridge by the same company – this carries Restalrig Road. Pronounced Ressle-rig, the name means *ridge of the miry land* – appropriate on my day! – from *lestal* a northern dialect term for mire and *rig* a linear field.

With Seafield now ¼ mile distant, the Restalrig Railway Path gradually curves westwards in a modest cutting. Pass by a tarmac path on the right. (This leads to Seafield Street, once the site of the Eastern General Hospital. An archway, with 'EGH' above, is a small remnant.) Continue for a few yards to the site of Seafield Junction, on the left of which is Seafield Cemetery and Crematorium. From the Junction, a freight line went straight ahead, crossing

The underbridge at Seafield Cemetery entrance. *Robin Howie*

Today's pedestrian bridge over Seafield Place. *Robin Howie*

Seafield Road (no sign now of the bridge) to enter the expansive Docks area. This merged with the track which still runs from Portobello West Junction on the Edinburgh-Berwick main line – in fact the sometime Leith arm of the Edinburgh & Dalkeith Railway, designated 'freight only' from 1905 and partly overlaid by the 'Lothian Coal Lines' from 1917. Some rationalisation of the two layouts, venerable Edinburgh & Dalkeith and Leith East newcomer, took place – and on the whole it was the North British Company who benefitted from the Caledonian's heavy outlay.

Continuing by the perimeter of the cemetery, pass by an old embankment on the right. (Heading east, this ends abruptly overlooking Seafield Street.) The Restalrig Railway Path (also signed 'Cycleway 10 and Leith Links ¼ mile') crosses three pedestrian bridges: the second leads high above the remnants of the huge abutments at the entrance to the cemetery and crematorium, while the third, opened in 2012, crosses Seafield Place whence the Restalrig Railway Path gradually descends to the east end of Leith Links. The line continued to Leith East goods yard but has been obliterated by industrial development, as has all trace of South Leith station, which dated from 1831 and the Edinburgh & Dalkeith Company's early expansion.

Dating back to 1813, an impressive building of classical style on the corner of Seafield Place is significant in being the city's only surviving building to have housed seawater baths – there were seventeen, of varying temperature – and also a swimming pool. People flocked to Seafield Baths, believing the seawater used could help cure illness and injury. The baths eventually closed in the early 20th century and the building has been converted to flats and offices.

Why not end the walk with a stroll through Leith Links, famous in the history of golf. Both Charles I and the future James II and VII were said to have played golf on the links while in residence at Holyrood Palace. The rules of golf developed in Leith were adopted by the Royal and Ancient Company of Golfers on their move to St Andrews in 1777.

WALK 7

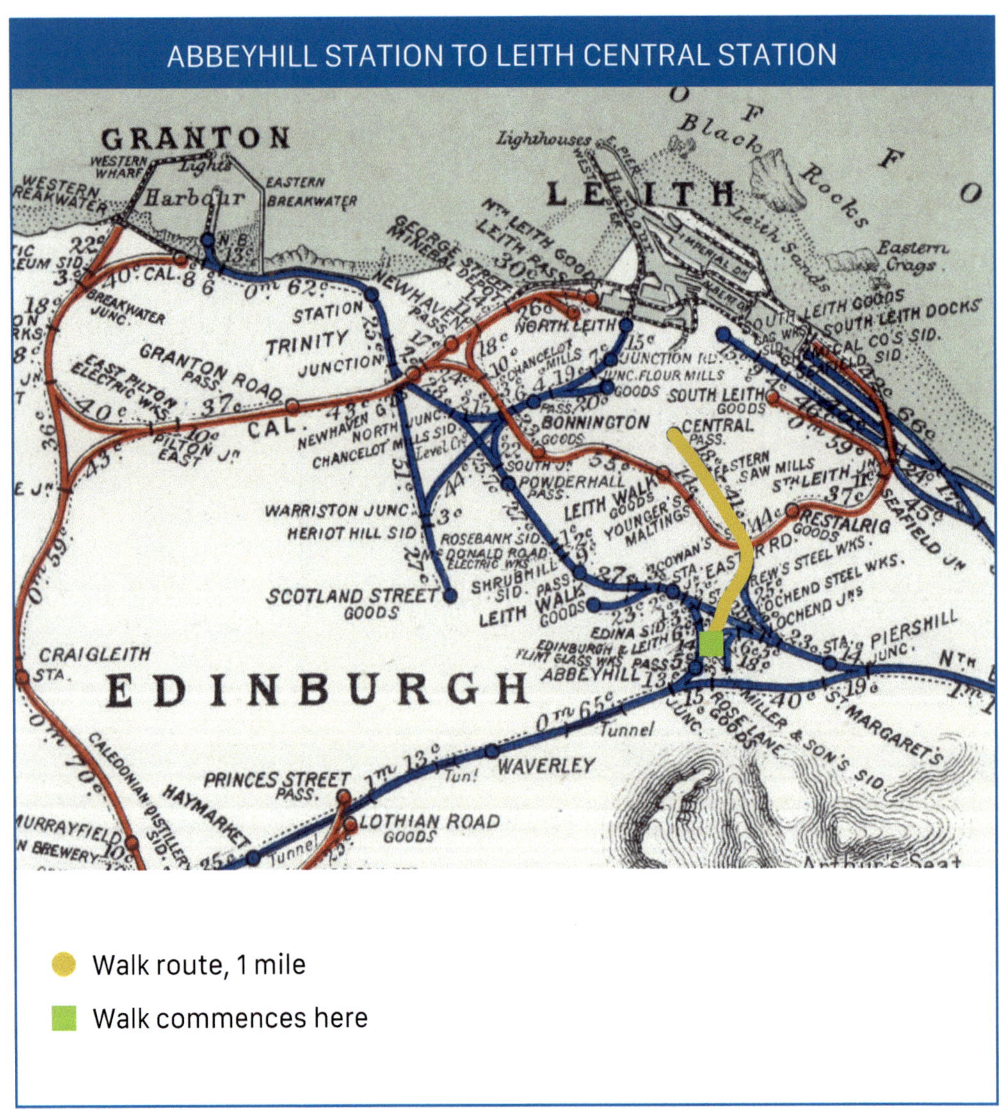

● Walk route, 1 mile

■ Walk commences here

In 1889 the Caledonian Company formed ambitious plans for a 'circle' round North Edinburgh – by extending their Newhaven branch into Leith, tunnelling under Calton Hill, tracing the length of George Street by cut-and-cover, and so back to Princes Street Station. Their plan had caused considerable alarm. Proprietors at the east end of Princes Street could recall the rumbling and tremors below their shops in the days of the Scotland Street tunnel, while the George Steet smoke vents would have been both noisome and obtrusive. Nevertheless, the town council gave the scheme its backing and the Caledonian persisted over the following decade, until Parliament eventually concluded that the benefits did not outweigh the negatives. The North British riposte – almost certainly what it was – saw the construction in 1903 of their expensive double-track branch from Abbeyhill to an imposing new terminus, styled Leith Central, at the foot of Leith Walk.

Of similar scale to the Caledonian's generously remodelled Princes Street Station, Leith Central was the largest station built (as opposed to extended or remodelled) in Britain during the 20th century.

Occupying a whole town block, bounded by Leith Walk, Easter Road and Duke Street, it symbolised the might of the North British and expressed their resolve to dominate Leith. They overinsured against a threat which was never to be repeated. There was no goods traffic,

A locomotive representative of the post-1948 British Railways Standard classes heads a train at Leith Central. The elevated four-platform terminus (closed in 1952) had a lofty overall roof with an elaborate end-screen. *NBRSG*

Derelict Leith Central station and signal box. The bridge over Easter Road has been removed but the site would not be fully cleared until the 1980s. *Roberton*

In 1904, Leith Corporation took over the Edinburgh cable trams within their jurisdiction and electrified their part of the system. Conversion is in progress. *Tom Wright*

but for some time local passenger business was heavy – thanks to the inconvenient divide between the Edinburgh tram system and its Leith counterpart.

The station, never used to capacity, closed in 1952 after a long decline; and thereafter the large train-shed served as a carriage depot. From 1957 it was equipped to service the diesel units newly introduced to the Edinburgh–Falkirk–Glasgow main line.

Leith's coat-of-arms

The upper structure of the station, and with it the diesel depot, was demolished in 1988. In Irvine Welsh's *Trainspotting*, his character Renton sadly reflects on the station closure, its barren, desolate vault soon to be demolished and replaced by a supermarket. An old drunkard sarcastically enquires – 'What yis up tae lads? Trainspotting, eh?'

The Abbeyhill–Lochend area is now smothered by new-built flats and by Meadowbank Retail Park, completely subsuming not only the start of the line from London Road Junction but also Lochend North Junction, from where a spur curved south-east to join the Piershill loop line.

The walking distance to Leith Central Station is only one mile, suggesting a short saunter of less than an hour, but be prepared for a half-day's outing, incorporating Lochend Park and historic Leith whose motto is 'Persevere'. Pick a good day, have faith in the weather forecast and enjoy 'Sunshine on Leith'.

Start from the Crawford pedestrian bridge, reached from Easter Road and the far end of Bothwell Street, as described in Walk 2: Waverley via Abbeyhill to Trinity Junction. In crossing the fenced-off line to Piershill Junction, note how the bridge gives access to the south side of Hibernian's Easter Road stadium – a short cut both for away supporters and for those tracing the route of the railway.

Follow Albion Terrace and turn right on Albion Road to where it is crossed by Lochend Butterfly Way. Turn right on the Way, the first definite legacy of the lost railway, and cross the remains of the Piershill loop by the Lochend Butterfly Bridge. On the far side is Lawrie Reilly Place – a nod to one of the 'Famous Five', the Hibernian forward line during the late 1940s and early 1950s … and, since you ask, I have remained a 'Hibs' fan since those schoolboy days.

Retrace steps to the Way and pass by a high brick building still lettered 'JAMES DUNBAR'. This soft-drinks factory, which closed in the 1970s, gave its name to the 'Dunbar End' of the football ground. The Way turns right then left, just below Easter Road stadium, to overlook the west side of Lochend Park. Buildings on the west side of the Way now occupy what was a broad and deep, ½ mile-long cutting in which stood the short-lived (1950–1967) Easter Road Park Halt, whence a steep path led to the stadium. This provision meant that a significant number of fans could travel by football 'specials' from Glasgow direct to the

Easter Road Park Halt, J. McEwan (Bill Lynn Collection)

Easter Road Park Halt, with the match programme of Saturday 8 April 1950. *Tom Wright*

stadium. With only a single wooden platform, the Halt lacked facilities for return rail travel and supporters were required to make their way to either Abbeyhill or Waverley. The Halt was officially opened before the game against Clyde on Saturday 8 April 1950, with the front page of the match programme picturing the cutting.

The immediate post-war period had seen games throughout the country watched by crowds in unprecedented numbers. A huge banking, extending the east side terracing, increased capacity, such that the 1950 New Year's Day derby between Hibs and Hearts attracted a crowd of 65,840 – Edinburgh's biggest ever football crowd. (The 149,269 who saw England defeat Scotland 2–1 at Hampden in 1939 remains to this day a record for any game played in Britain.) However, the attendance boom quickly became a thing of the past. (Post-war cinema attendance shows a similar pattern.) In the 1983–84 close season the size of the terracing was reduced; then it was completely demolished (2010) to make way for the new east stand of smaller capacity.

Following this demolition, thousands of tons of waste were very efficiently transported to in-fill the adjacent cutting – a deep burial ground for the Halt, which, given present-day landscaping, is now difficult to imagine.

Leave the Way for an anti-clockwise walk round Lochend Park and the historic Lochend Loch, home to a wide variety of wildfowl – swans, geese, coots, moorhens, mallards and herons. On the far side, pass below Lochend House, built around 1820, which incorporates the remaining gable of Lochend Castle, a 16th-century L-plan tower-house built high on a precipitous rock face. The castle was home to the Logans of Restalrig (see also Walk 6: Newhaven Station to Leith East).

Lochend Loch *Margaret Howie*

Lochend 'doocot' (pigeon house)
Robin Howie

Sir Robert Logan played a notorious part in the Gowrie Conspiracy, an unsuccessful and never satisfactorily explained plot to kidnap or assassinate James VI in 1600. (A lost play about this might-have-been regicide helped shape Shakespeare's *Macbeth*.) Latterly used as a children's centre until a fire burnt out the roof, it is now, following rebuilding, in private ownership.

On the north-west side of the loch is a 16th century dovecot ('doocot'). Built to supply pigeons to Lochend Castle, it is a beehive structure. There is evidence of a chimney, thought to have been used as an incinerator to burn infected clothes during the plague of 1645, when, according to parish records, over 56% of the population of South Leith was wiped out. Old maps refer to the structure as the 'plague kiln' but it now provides a haven for feral pigeons.

Fed by underground springs (there are no streams flowing in or out), Lochend Loch wasfrom around 1650 the main water supply for the Port of Leith. An 18th-century pump-house, between the doocot and the loch, facilitated the process, which earlier had relied on gravity. The final water supplies from the loch were sealed off around 1920. Kirkwood's 1817 map of Edinburgh shows the doocot almost on the shore. Nowadays the loch has been partially filled in for safety reasons and its depth much reduced. Drainage is through two overflow pipes; however, the topography means that in wet conditions the pond can occasionally overflow, flooding the perimeter path – an overflow all too obvious on the day of my visit.

The carpark where Leith Central once stood. *Robin Howie*

Return to the Way which heads north following the line of the cutting. Cross Hawkhill Avenue diagonally then, via a steep downhill zigzag path, join the Restalrig Railway Path which provides the concluding stretch of this Walk.

Beyond a fenced-off area, the sports grounds of Leith Academy have subsumed the railway, but a convenient detour is to hand. Turn left on the Restalrig Railway Path (signed 'Foot of Leith Walk 2/3 mile') which, curves north-west past the new Leith Academy to reach lower Easter Road. Turn right as signed ('Foot of Leith Walk ⅓ mile') and, just before Easter Road meets Lochend Road and Duke Street, note the Tesco Superstore car park on the left – the site of Leith Central Station.

At this point the tenements on the east side of Easter Road show an odd change of architectural style from traditional to modern and back again. The obvious modern section, a four-storey block, fills the gap created by the North British Company in demolishing those properties so as to give access to the station – testimony to the scale of the project. A four-track bridge, of which nothing now remains, carried the railway over Easter Road. With the advent of double-decker buses, the relatively low height of the bridge together with a likewise low bridge at Abbeyhill, meant that only single-deckers could operate on Edinburgh's time-honoured No.1 bus route.

The station's four platforms and adjacent sidings under a steel-framed over-all roof, stood some 15 feet above street level: this was dictated by height of the Easter Road bridge, with

An early view of Leith Central bounded by Duke Street and Leith Walk.

Old Leith Academy. *Robin Howie*

insufficient distance thereafter in which to take the railway to ground level. The castle-like remnant of a massive wall on the right of the car park is a reminder of the building's scale.

Continue to the far end of the car park and pass by Wonder World (a Soft Play area for children) and then by the Superstore to reach Leith Walk. Turn right and pass by the Central Bar, today's almost solitary relic of the station's ground-level retail premises – a down-to-earth, late-Victorian, drinkers' pub with one of Scotland's most stunning public-house interiors. Entrance porches (with mosaic flooring and stained-glass windows) lead into a room which has ceiling-to-floor tiling. The side walls are notable for four tiled panels of sporting scenes, with tall, narrow mirrors between. There was once a direct door connecting station and pub.

Continue to the corner of Duke Street, where stairs led to the ticket office and waiting rooms at platform level. (From almost underneath the Easter Road Bridge there was an additional entrance to the platforms by two stairways, in order that passengers need not make the long detour along Duke Street to Leith Walk to catch their train.)

Continue along Duke Street – a stretch where the upper storey of the station offices survives. At the southwest corner of Leith Links is the handsome old Leith Academy, now converted to flats.

WALK 8

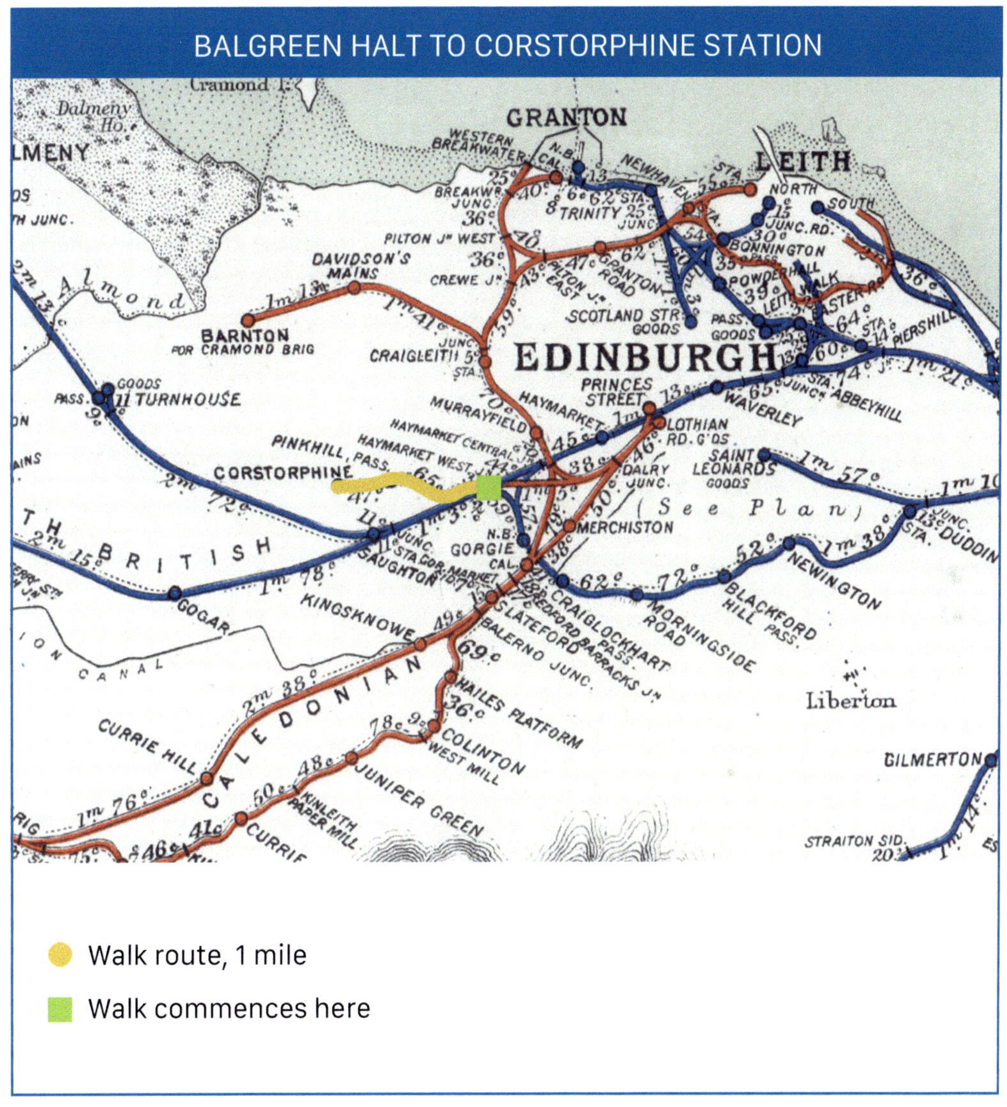

● Walk route, 1 mile

■ Walk commences here

The earliest recorded name of Corstorphine was *Crostorfin*. This possibly means 'Torfin's crossing'. In ancient times, much of the land in the area consisted of small lochs and marshes, with Corstorphine situated at the best crossing point. The identity of Torphin is not known for certain, but he is likely to have been a local baron who commanded a stronghold by the crossing.

Surprisingly, the small village of Corstorphine had a station as far back as 1842, albeit situated one mile to the south on the Edinburgh & Glasgow Railway, just east from the foot of today's Saughton Road North. The station's name was changed to Saughton in 1902 when the North British Company's Corstorphine branch opened. Edinburgh's approach line to the Forth Bridge diverged at the new Saughton Junction (1890). Four-tracking from Abbeyhill followed, through the rebuilt and much enlarged Waverley Station, to accommodate increasing traffic, so that the original 'Corstorphine' became four-platform. As 'Saughton', it closed temporarily in 1917 – a wartime economy measure; final closure came in 1921.

Commencing west of Haymarket and running by Balgreen and Pinkhill, the branch terminated on Station Road, at the east end of Corstorphine High Street. The journey to Waverley took only 14 minutes, allowing some commuters to return home for lunch; but, in common with other late inner-suburban additions, the line came to face tram competition.

Balgreen Halt, where the Corstorphine branch commenced. Note the four-track main line. The signals in the middle distance mark Haymarket West Junction, which gave access both to Caledonian Princes Street and to 'The Sub'. Stevenson, courtesy *NBRSG*

For operating convenience, over the years certain services extended to Musselburgh, North Berwick or Gorebridge, while Haymarket depot's express locomotives, beginning or coming off duty, were sometimes given Corstorphine 'turns', which reduced engine movements on the busy tracks between Waverley and Haymarket. Well into the days of British Railways, a local stopping service to Berwick started from Corstorphine and its coaches were serviced there.

Use the pedestrian crossing on Balgreen Road to reach a broad tarmac ramp (signed 'Balgreen Tram Stop' and 'Corstorphine 1 mile'), which leads to the modern tramway paralleling the extant main lines – six tracks if the tram lines are included. Today's Balgreen tram stop is adjacent to the site of the abandoned Balgreen Halt.

How the 'wheels of time' go round! Overlooking tramway and railway is the splendid 1925 Jenners Depository and Furniture Store, designed for easy horse-and-cart access. (At that date the motor lorry had by no means triumphed completely.) The distinctive red building is a useful landmark for eastbound ScotRail passengers who intend to alight at Haymarket Station.

Beyond the tarmac path divides. Curving right is the branch embankment, on the left of which are allotments and then two buildings. The second building is the clubhouse of Carrick Knowe City of Edinburgh public golf course, from where there are glorious views, with Corstorphine Hill as a backdrop. The clubhouse address is 27 Glendevon Park – vehicle access is from that road and under the former railway. The embankment path, which skirts Pinkhill, forms part of the John Muir Way (Helensburgh-to-Dunbar).

Jenners 'landmark' Depository and Furniture Store and today's Balgreen tram stop.
Robin Howie

Carrick Knowe golf course. *Robin Howie*

The considerable remains of Pinkhill station straddle the path. It was built on a curve with the Up platform, on which stood a waiting room, slightly longer and higher. By the broad overbridge is a white building, once the booking office and now a very small garage. There was no goods yard. Pinkhill had platform signs 'Alight here for Zoological Park' - Edinburgh's Zoo, opened by the Royal Zoological Society of Scotland in 1913. Past the station site and beyond the Forestry Commission Headquarters building, a sign on the right points to the Zoo, ¼ mile distant, which today receives over 600,000 visitors a year with penguins a regular attraction. Two giant pandas from China were 11 years 'in residence' and became celebrities.

The substantial remains of Pinkhill station. *Robin Howie*

Corstorphine station where a 'scissors' crossover (today a rare railway feature) facilitated the release and 'run round' of arriving locomotives. *Sanderson collection, courtesy NBRSG*

The tarmac way comes to a sudden end, though a path continues to the site of Corstorphine station. The building, platforms and sidings have been subsumed by the Paddockholm housing development. Erected in 1919 in front of the station building, the Corstorphine War Memorial was re-sited in 1928 and now stands opposite the library in Kirk Loan. 'Station Road' is the sole reminder of the railway's 65 years' presence.

The station had very long platforms, which allowed mainline trains to be washed and cleaned there, and the point-work, for release and run-round, could accommodate main-line engines.

The Northern Belle, the London & North Eastern Railway's pioneering 'hotel-train' of the 1930s, offered a week-long 'cruise' of that company's system at a cost affordable by those on a middling income. It combined 'day' and 'night' vehicles, and the latter were regularly stabled at relatively peaceful Corstorphine. The title has been revived by one of today's luxury-train operators, who market a more exclusive and much more expensive railway experience.

WALK 9

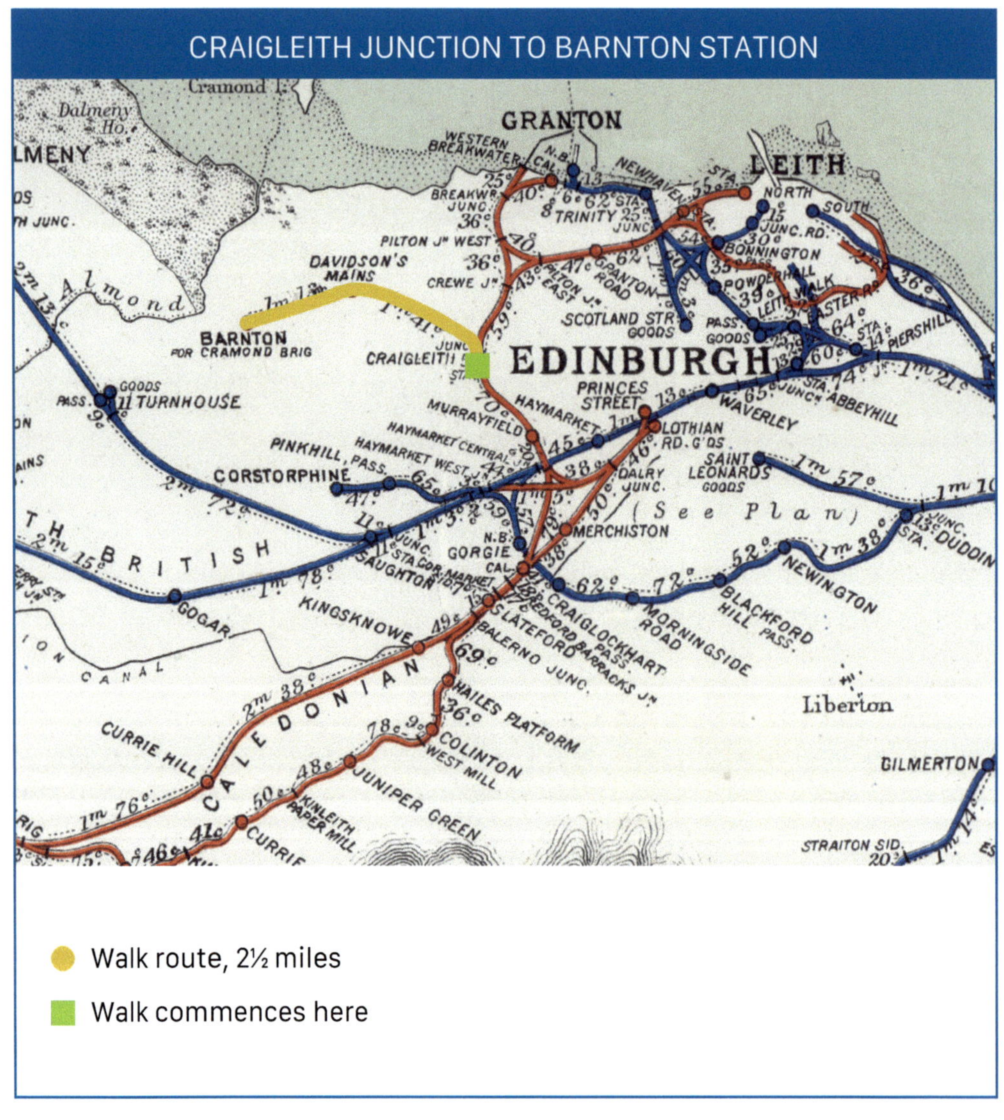

● Walk route, 2½ miles

Walk commences here

The 2½-mile, single-track Caledonian branch (1894) from Craigleith Junction had a station at Davidson's Mains (originally Barnton Gate), with a yard and coal depot, and terminated at Barnton (originally, Cramond Brig). The company looked to open up new residential districts west of the then city – thus pre-empting similar North British ambitions – and the venture proved worthwhile. The line might have become a 'circle', returning to Edinburgh via Corstorphine, but it was the North British who built the branch line to Corstorphine (see Walk 8: Balgreen Halt to Corstorphine). A more ambitious western 'circle' from Barnton to Balerno, via Ratho, was also briefly considered.

This Walk is a game of two halves. Initially a tree-fringed ascent on the Blackhall Path, the stroll to Davidson's Mains (known as Muttonhole until about 1850 – the provenance of the name is uncertain) is notable for four structures – the less-than-attractive Telford Road (A90) bridge; a fine masonry arched bridge by the site of the House o'Hill Halt (1937–1951); then those under Main Street and Silverknowes Road. (The last has an 1892 plaque.) No evidence remains of the LMS Halt near the summit of the line at the northern end of the House o'Hill district.

ABOVE: Craigleith Junction survives after a fashion, for walkers and cyclists at least, some 70 years after the passenger service to Barnton ceased. *Robin Howie*

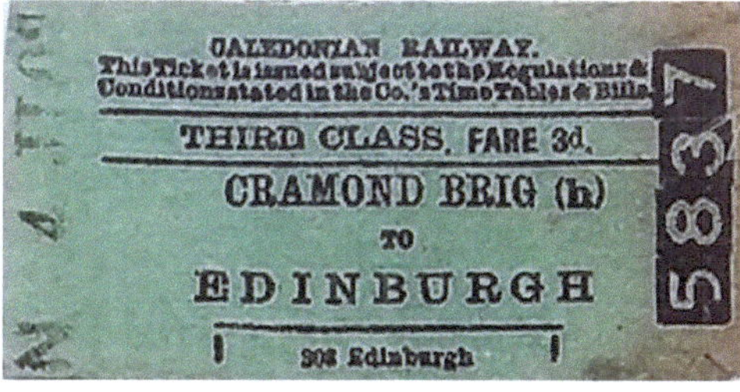

LEFT: Cramond Brig station (1894) became Barnton from 1903.

Blackhall Path today, where a handsome overbridge proclaims its railway ancestry. *Robin Howie*

The approximate site of House O'Hill Halt, an LMS addition to the Barnton line. *Roberton*

The approximate site of Davidsons Mains station. *Panton via Railscot*

Steps lead to the masonry bridge just mentioned, which once carried traffic but is now pedestrian-only. Pass under Main Street whence the cycle route curves right. However, continue curving left under Silverknowes Road on what appears at first to be rough tarmac but becomes a wide, often muddy, path with ballast underfoot. Enjoy this rare survival – a few yards of old track bed. Continue to the Tesco Supermarket and car park at Davidson's Mains, which have left no trace whatever of the sometime station facilities.

Although the second half of the walk may be disappointing, there is one final gem to be discovered.

The accommodation bridge at Barnton, with its arch filled in. *Robin Howie*

The railway took a straight line to Barnton station, situated on Whitehouse Road, just north of the Barnton roundabout – an approach now subsumed by houses. The station site and goods yard are occupied by a small shopping complex and no immediate evidence of the terminus can be identified. Reach the site using a scenic route which, following Barnton Avenue and Barnton Avenue West, passes between two golf courses (Royal Burgess and Bruntsfield Links) to reach Whitehouse Road, where turn left to the shopping complex. (The alternative route via noisy Queensferry Road is not recommended.) On arrival, walk up Barnton Grove. On the left is Barnton Court, one of the many blocks of modern flats, and at the rear, adjacent to the Royal Burgess golf course, is the arch of an accommodation overbridge, bizarrely stranded in such domestic surroundings. The eastern face of the overbridge can be seen by continuing along Barnton Park View.

WALK 10

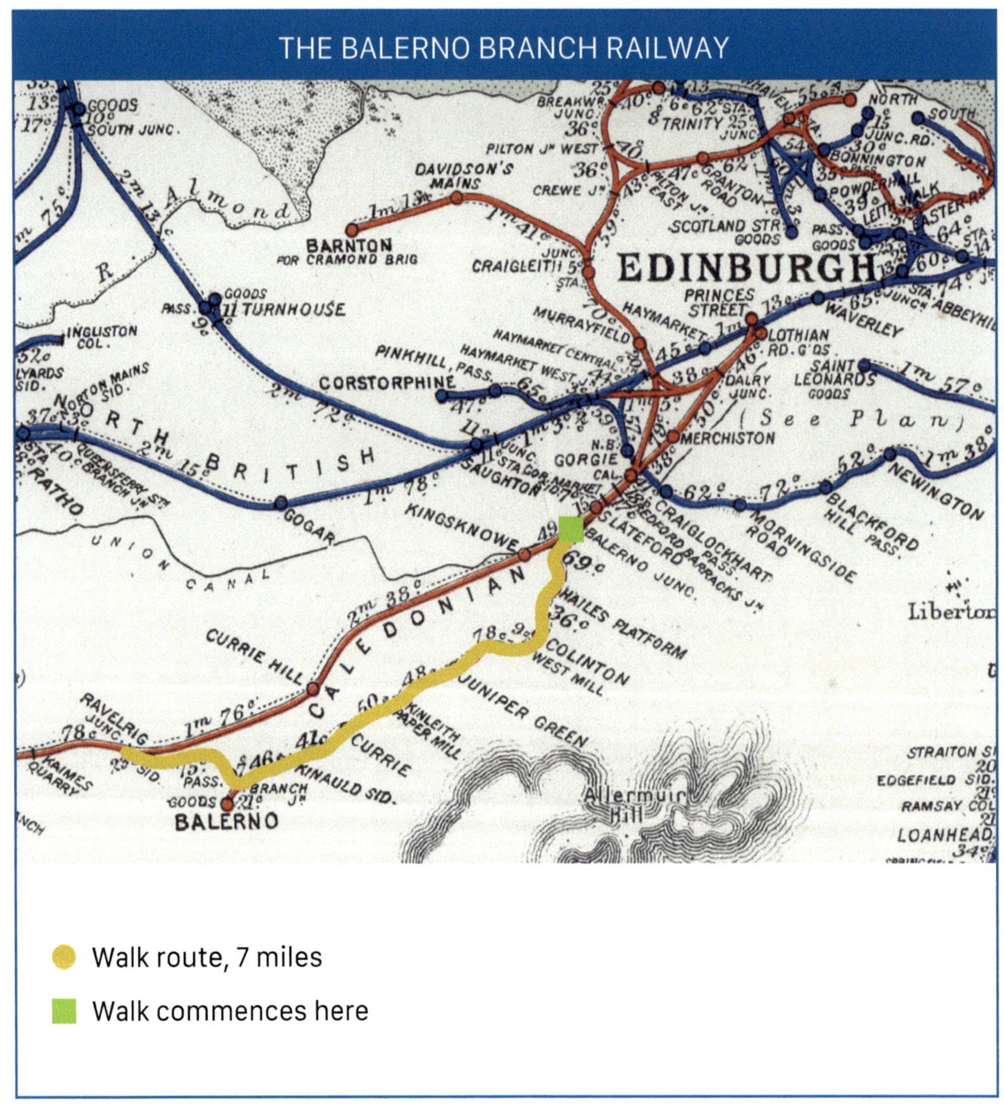

● Walk route, 7 miles

■ Walk commences here

The Water of Leith (initially the West Burn) rises from the Colzium Springs on the northern slopes of the Pentlands and flows via Harperrig Reservoir to its confluence with the Bavelaw Burn at the northern end of Balerno. Thence through Edinburgh to its estuary at Leith, it is accompanied for 12¾ miles by the Water of Leith Walkway, first completed in 2002 and now entirely macadamised. The 5½ miles from Slateford to Balerno was for some years *literally* a trackbed walk.

Lengthy sections of the excellent Walkway give the opportunity for secluded and pretty walks that at times could be in the countryside, although the Water bisects the city.

In following the curving stream, crossed four times, considerable construction works – bridges, cuttings, embankments and a tunnel – were required in order that the railway could follow the winding stream. The overall ascent is 300 ft and locomotives were specially designed for this relatively testing duty.

The 'branch' was in fact a loop from Balerno Junction, just west of Slateford, rejoining the Caledonian Company's Edinburgh-Carstairs route at Ravelrig Junction, one mile north-west of Balerno. The single-track line was built primarily to service the many water-powered mills which through the centuries produced a wide range of products. Little evidence remains of all this industry, and housing has replaced the mills.

Passenger stations were provided at Hailes Halt, Colinton, Juniper Green, Currie and Balerno, which encouraged expansion of the mill villages along the way, gradually extending the city's suburbs aalong Lanark Road. The line opened for all traffic on 1 August 1874. The first passenger trains, the 7.50 am from West Calder (on the Shotts cut off) to Edinburgh via Ravelston Junction and Balerno and the 8.00am Edinburgh to Mid Calder via Balerno and Ravelston Junction, crossed at Currie Station, initially the only passing place.

By the early 1920s, given its convenient access to several golf courses – Hailes for Kingsknowe, Colinton for Torphin and Juniper Green for Baberton – the railway was sometimes referred to as the Golfers Line. Baberton even had a bell installed in the clubhouse to give 10 minutes warning of the departure. In similar vein, the North British cultivated golfing traffic between the city and their East Lothian stations.

WATER OF LEITH VISITOR CENTRE TO COLINTON STATION 1½ MILES

The Balerno Railway left the main line ¼ mile west of Slateford station. Close by today is the Water of Leith Centre on Lanark Road, an appropriate start to the Walk. The Water at this point is crossed by two arched masonry viaducts, one carrying the 1822 Union Canal and the other the Caledonian main line (1848). Stone for the 14-arch railway structure, built entirely by manual labour, came from several local quarries which the canal accessed.

Hailes Quarry has an interesting history. At peak activity, it employed 150 men and 100,000 tons of stone were extracted annually. The quarry was abandoned in 1902 when it had reached a depth of 100 ft and become flooded. In 1949 water was pumped out into the Murrayburn to make way for a landfill site, grassed over; and in 1981 this was made into a public park worthy of a visit. It is rumoured that an elephant from Edinburgh Zoo is buried there – such rumours usually have *some* basis though I am unaware of any attempt to test the story. However, the tale of a horse entombed in one of the hollow piers of Glenfinnan viaduct on the West Highland Railway has been proved more-or-less true. Investigation has has shown (unless there was a second horse, and a similar incident!) that the skeleton lies within Loch-nan-Uamh viaduct, further to the west.

The Union Canal aqueduct with the railway viaduct carrying the main line behind. *Robin Howie*

The Water of Leith Conservation Trust – the first river charity to be established in Scotland – works to preserve and enhance the Water, its heritage and wildlife; their Visitor Centre promotes education and facilitates recreation, and they work with volunteers and community groups to deliver over 200 river clean-ups and habitat improvement drives annually. Pay a visit to the Centre – your support will be appreciated. Scottish Water is currently protecting the environment along the valley and improving the waste-water network by upgrading the overflow chambers, 'which help to prevent items, wrongly flushed down the toilet, from reaching the river'.

On my visiting days there were a couple of temporary, but well-signed, detours from the trackbed.

From the Visitor Centre, head west on Lanark Road to where it is crossed by a pedestrian bridge, reached by a ramp on the north side. For a short detour, cross the Union Canal by the former railway bridge and continue via the short embankment to a high fence by the main line, where all trace of Balerno Junction has gone. Return to the pedestrian bridge, signed 'Colinton Dell ½ mile' and 'Colinton 1 mile', and cross Lanark Road.

Welcome to the former railway, which passes under Redhall Bank Road (part of the original Lanark Road, later shifted northwards to allow more quarrying) and over Bogsmill Road. Here begin the embankments and cuttings which took the sinuous line high above the wooded gorge of the Water – a haven for wildlife, as it is to some extent all the way to

Colinton tunnel, looking west. The mural recalls the Caledonian Company's blue-liveried locomotives. (North British 'bronze-green' was subdued by comparison.) *Robin Howie*

Balerno. It is no surprise to learn that Colinton (originally Colintoun) means 'village of the woods.'

Continue to what was once the site of Hailes Halt, overlooking Katesmill Bridge and a weir. For a pleasing detour, follow the stepped path, signed 'Colinton via Katesmill', and descend to the Water. The bridge here is a girder span but, as parts of the wooden decking are in disrepair, access is barred. The weir is best appreciated during a wet spell. Return to the course of the railway from where a path on the right climbs to Lanark Road. Higher up is a sign 'Knockifulbraehead' – quite a mouthful!

Continue to Colinton tunnel 150 yards long; thanks to the tight curve, it seems longer. (High above, Spylaw Bank Road descends to Dell Road but the exact spot where road crosses railway is difficult to judge.) The tunnel retains its 'echo qualities' but is no longer a dark and scary place. The Colinton Tunnel Murals – a stunning visual celebration, and with no anti-social graffiti – link the story of the railway to the history of the community.

The words on the wall quote Robert Louis Stevenson's poem 'From a Railway Carriage', a lovely evocation of a child's first train journey:

> 'Faster than fairies, faster than witches,
> Bridges and houses, hedges and ditches; …
> And here is a mill and there is a river:
> Each a glimpse and gone for ever!'

Which contrasts with his alarming description of the Scotland Street Tunnel (see Walk 1: Waverley Station via Scotland Street to Granton).

At the west end of the tunnel is the site of Colinton Station, spanned by Bridge Road. All that remains is the access road, descending to a small car park.

COLINTON TO JUNIPER GREEN STATION 1¼ MILES

With Juniper Green 1 mile distant, head beyond Bridge Road for Spylaw Park, where in the late 18th century James Gillespie had a prosperous snuff-mill. (With his legacy, James Gillespie's High School was founded in 1803.) Adjacent to the mill, he built an imposing mansion house, now a listed building and redeveloped into private flats. Part of the mill can still be seen at the back of the house.

Beyond the station site, the former railway has its first crossing of the Water, to what is now a housing development close to West Mill Road on the south bank. Here was the West Mills site (1909–89) of A&R Scott Ltd, producers of Scott's Porage Oats. Milling now takes place at a modern facility at Cupar, the largest oat mill in Western Europe.

From a brief spell on the access road to the West Mill housing development, the macadamised trackbed is regained – a wide corridor fringed by trees. Then back to the north bank by a fine masonry bridge, it curves left to pass a new pedestrian bridge over the Water, whence a path burrows beneath the old trackbed and a stepped way climbs to Lanark Road.

Beyond is the all-too-obvious, utilitarian and very high A720 City Bypass Bridge spanning the valley. The noise from the weir just below the Bridge is not intrusive, whereas the traffic noise from the bypass certainly is! Speed on!

The footbridge below the A720 Edinburgh Bypass. *Robin Howie*

Downstream from the former Colinton station. *Robin Howie*

The remains of Juniper Green station when goods trains still ran. *Shaw via RailScot*

Ignoring the path to the right which ascends to Juniper Green, bear left by the Water to avoid the housing development occupying the sometime formation. On the right is a waterwheel, a reminder of the past; here an original redbrick warehouse has been converted into town houses. Beyond is the site of Juniper Green Station, now landscaped to leave no trace.

Juniper Green was earlier called 'Curriemuirend' – appropriate given that it had not then grown into a residential outpost in its own right. For a brief visit to the village, head up Baberton Loan, the station access road. On the far side of Lanark Road is the Railway Inn (established in 1891), the Railway Barbers of more recent vintage, and Molly's Delicatessen and Cafe, recommended for refreshment. Opposite Molly's, a narrow, stepped lane (signed 'Currie 1½ miles') leads back to the sometime formation where another pedestrian bridge spans the Water, giving access to Woodhall Road.

JUNIPER GREEN TO CURRIE STATION 1¼ MILES

After a short stretch where the railway was squeezed between the Water and a high embankment, recross to the south bank and pass under the arched masonry bridge carrying Blinkbonny Road. Pass what is in effect a small modern village located in a large open area. This was the site of Kinleith Mill, the biggest mill on the Water of Leith – it was demolished in 1966. (Vehicle access to Lanark Road is by a new bridge over the river to more houses and then to Blinkbonny Road.) Currie is now ½ mile distant. The open area quickly changes to a wooded glen, where, above a weir, the Water briefly appears canalised.

Reach Kirkgate, the access road for Currie Kirk and the cluster of buildings which represent the original village. The topography has ensured that this historic core, including the Water of Leith, has remained undeveloped; it is now designated a conservation area. There is no accepted derivation of the name 'Currie' but it is possibly from the Gaelic *curragh/ curagh* – a wet or boggy plain.

Descend by steps to Kirkgate and follow a path downstream by the Water's edge. On the right, amid a wet area, is St Mungo's Well through which flows a tiny stream. The stone basin with side walls is undistinguished, but until 1872 the well supplied drinking water for the village and its 250 residents.

St Mungo's Well. *Robin Howie*

The site of Currie station. *Robin Howie*

Cross the Kirkgate bridge to reach the site of Currie station. On the left is the stationmaster's house, set within a lovely garden, and an adjacent row of cottages. On the right are a few remnants of a wooden platform; but it is hard to visualise the station, set on an embankment high above the Water, with its passing loop between side-platforms and a ticket office and waiting room on the northern side – altogether a tight fit. No evidence of the station building remains. However, the goods yard on the south side is easy to determine. The large brick-built goods shed has been converted to a private dwelling and the area landscaped. The yard (still with traces of the loading bank) was accessed from the west side of the station. Balerno is now 1½ miles ahead.

CURRIE TO BALERNO STATION 1½ MILES

Note the old telegraph poles on the right – once the accompaniment of any railway journey. Back in the 19th century the railway system and the telegraph system expanded together (Charles Dickens described the lineside telegraph wires as music scores ruled across the sky). On the far bank are Currie Bowling Club and, later, a large complex of new riverside apartments styled 'River Mill'. Beyond this spot, the line again crossed the Water to reach Waulkmill Loan on the north bank. Waulking' is a hand-finishing process which shrinks 'white' (undyed) cloth by binding warp and weft more tightly, but water wheels, driving trip hammers, were employed to 'full' (felt) the cloth centuries before spinning and weaving were mechanized – hence 'fulling mill'.

With Balerno village now close to hand, farmland and then Malleny Park (home of Currie Rugby Football Club) add a verdant touch to the south side of the Water. On the north side, the former railway is confined by a 15-feet-high 'Criblock' retaining wall below Lanark Road. Criblock is a gravity-type retaining wall – in this instance, pre-cast concrete cells back-filled with stone, giving the mass its stability yet allowing water to escape from the compacted material through the structure's open grid. This stops any build-up of hydrostatic pressure behind the wall and obviates its possible failure – of extreme importance given the wide and busy road immediately above. (Similar but smaller Criblock sections can be seen further downstream.)

The Water of Leith Walkway comes to an end just south of the Lanark Road/Bridge Road junction – a peaceful conclusion, despite the traffic but what about the railway? From the mid-point of the Criblock retaining wall, it curved sharply right and passed under Lanark Road (no trace now of the start of the short tunnel) to a cutting – the site of Balerno Station, somewhat distant from the old village.

Today's walker should use the path at the east end of the Criblock wall and cross Lanark Road, on the far side of which is the stone-built stationmaster's house, currently used for commercial purposes. Today's Station Loan, a private road further west, was not the station access. A stepped path descended to the single platform in the cutting, partially filled-in, where a house of three flats now stands. The lowest floor is below street level. From the street only the far end of the tunnel can be detected.

Most trains terminated at Balerno; but until a passing loop was installed they went forward empty to Ravelrig Junction on the main line, where locomotives could run round.

After the First World War, outer-suburban house building began in earnest. Recent developments – and more are planned – testify to residential pressures, as does the constantly busy Lanark Road. (Currie has a nearby station, Curriehill, served by Edinburgh–Shotts–Glasgow trains, but alas Balerno has not.)

Balerno had its own goods yard, separate from the station, now the site of Balerno Community High School. There were early days attempts by villagers to have their passenger

Balerno, in Caledonian Railway condition; the station acquired a passing loop but not an additional platform. *Currie & District Local History Society*

facilities resited. People going to the station from the village had to walk up Bridge Road and cross Lanark Road. A bad accident in 1924 involved passengers who had just alighted from an Edinburgh train. The steering on a bus coming from the city failed, and it ran into them as they waited to cross the road. Two girls from the village were killed and one man seriously injured.

For a detour, enter old Balerno, whose name derives from *Baile Airneach*, meaning 'townland of the hawthorns'. Turn left on Bavelaw Road to the entrance to Malleny Park and Gardens. The latter is renowned for its peaceful atmosphere and beautiful surroundings. Through a decorated wrought-iron gate, the walled garden opens out to the Four Evangelists (huge 400-year-old clipped yew trees), Victorian greenhouses and heritage rose plantings. Just outside the garden is a 'doocot' (pigeon house) with an unusual saddle-backed roof.

John Galloway's Balerno Bank Paper Mill, the largest in Balerno, was located near the centre of the old village. It closed in the early 1980s.

BALERNO TO RAVELRIG JUNCTION 1¼ MILES

Immediately beyond the station, the cutting has become part of private gardens, so follow the lane on the left to reach Dalmahoy Crescent. A grassy area on the right gives a further view of this deep earthwork. Continue through a large housing area to the end of the Crescent. At a T-junction turn right and steeply descend on what is known as Burn's Brae to reach where the railway bridged Ravelrig Road.

The central span has been demolished though the embankment on both sides is obvious. The short stretch leading back to the cutting is scarcely worth attention. However, heading west, the ¾ mile to Ravelrig Junction and the main line is most certainly worthwhile.

Climb to the embankment on which a lovely path with old ballast underfoot winds between the trees. The trackbed gradually slants towards the working railway – the latter is almost unseen, so you may be startled by a passing train! Joyous news, a real non-city exploration, in countryside which compares with parts of rural East Lothian. Though new houses may soon occupy the field on the left, the old railway will remain a protected path. The embankment comes to an end, followed by easy walking on the wide grassy trackbed, albeit one part might be soggy.

Cross a fence to enter a densely over-grown wood, then cut up to the left to a line of upright sleepers and traces of an earthen platform – the loading point for rock bogeyed from quarries on Ravelrig Hill and forwarded by rail. The former Ravelrig Junction is marked by a wider area some 100 yards to the west. Almost hidden in the wood are the ruins of railway workers' cottages – in this remote location they had only quarrymen for company.

Quarrying ceased in the 1930s but has resumed in recent times. Do take time to explore the lower slopes where traces of the old bogie lines can still be detected. Brake wagons controlled the gravity descent and horses pulled the empty wagons back uphill. Access is easy, by a worn path to a metal gate just above the dense wood.

Ravelrig Halt was situated by Dalmahoy Mains, whence a farm track still crosses the main line. Opened in 1884

Leaving Balerno. *Robin Howie*

Ruined cottages at Ravelrig Junction. *Morag Dunbar*

The site of Ravelrig Halt. *Morag Dunbar*

with two platforms, the Halt saw little use until Dalmahoy Golf Club was established in 1927, after which a sign 'Ravelrig Platform for Dalmahoy G.C' was displayed. No evidence remains.

With no ready access by road, Ravelrig Junction served on more than one occasion during the Second World War as a 'secret' (or at least secluded) stabling point for the Royal Train.

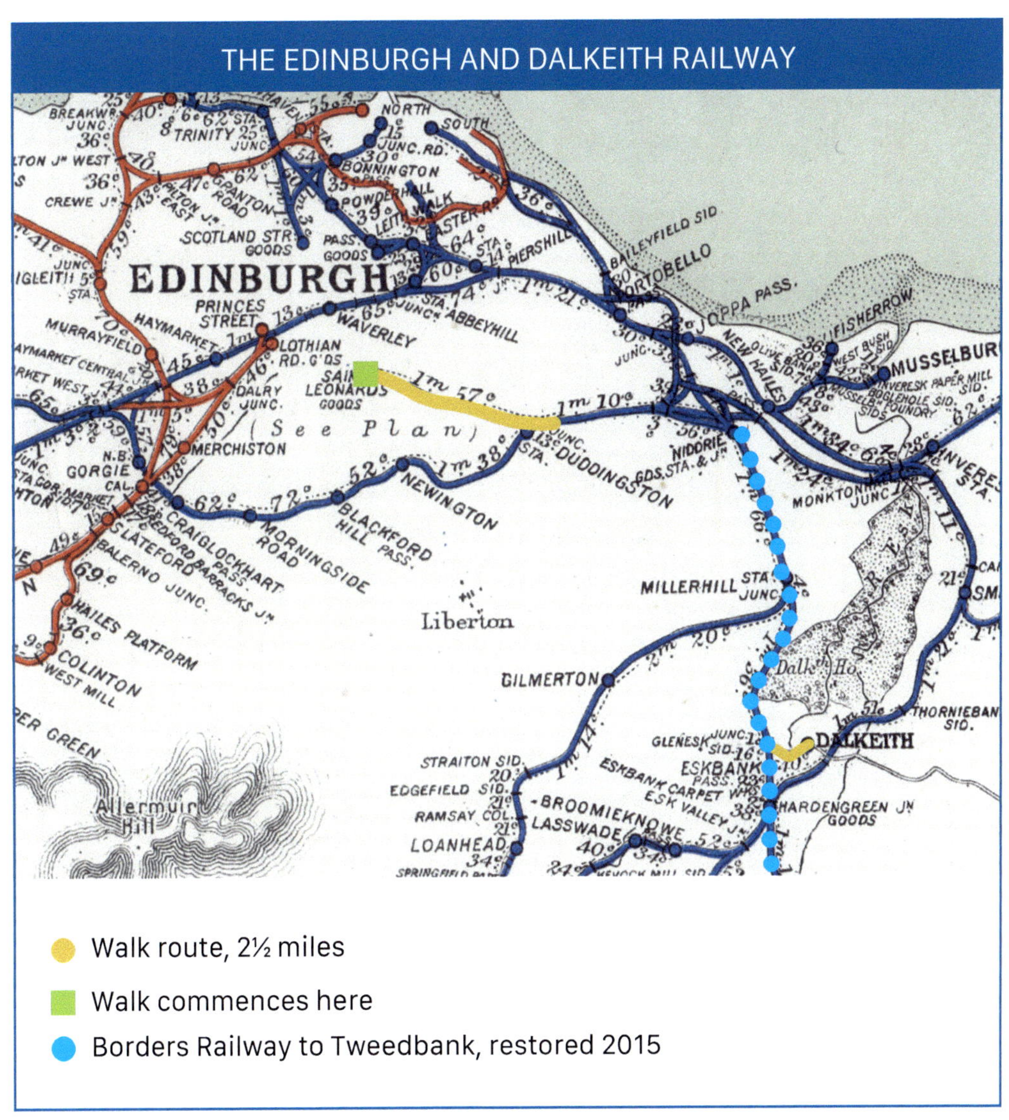

- Walk route, 2½ miles
- Walk commences here
- Borders Railway to Tweedbank, restored 2015

One of the earliest British railways and the first into Edinburgh was the 8¼ mile Edinburgh & Dalkeith, laid out for horse-haulage and opened in stages from 1831. It terminated at St Leonard's Station and Coal Depot. The approach to St Leonard's had a gradient of 1 in 30, dictating cable operation, and passed through a 572 yard tunnel completed in 1829 – one of the earliest railway tunnels in the world. Initially passengers were carried in wagons, open carriages and converted stage coaches (i.e. coach bodies mounted on a flanged-wheel chassis). With some 200,000 to 300,000 passengers each year, and more passengers per mile than the celebrated Liverpool & Manchester Railway (1830). Regular passenger traffic was diverted from St Leonard's after 1842.

The route became known as the Innocent Railway, perhaps in opening up the 'innocent' or clean countryside to dwellers in Auld Reekie, or perhaps because the slow speed of the horses meant that no intermediate stations were needed and passengers could get on and off where they liked, and with few accidents. It has also been suggested that the name 'innocent' derives from the fact that the company did not issue tickets to travellers in the early years, because passengers could not, or would not, make up their minds as to their destination. The rules of railway working were still being learned, and passenger business was at first leased to an independent contractor.

Following construction of a 70-ft-high wooden bridge across the valley of the North Esk in 1838, the line was extended into Dalkeith which lies on a ridge between the Rivers North Esk and South Esk.

By 1847 the track gauge had been changed to Stephenson's standard 4 ft 8½ inches, while horses had been replaced by steam locomotives. Now owned by the North British Company, the Edinburgh & Dalkeith was realigned, cutting off Newton village. The wooden Glenesk bridge was replaced by a masonry structure.

St Leonard's closed to passengers in 1860, though it continued as a mineral yard to 1968. In the last years of steam, the St Leonard's turn was accounted a relatively easy duty, appropriate for a driver nearing retirement. There was allegedly the bonus of an unofficial nip of whisky, in the course of shunting the yard. Much of the original line to Glenesk became part of the Edinburgh & Hawick Railway, later extended to Carlisle and celebrated as the Waverley Route by the North British Company. Closed in 1969, this has now been reinstated (as far as Tweedbank) as the Borders Railway; but little is left of the Edinburgh & Dalkeith – just the initial stretch from St Leonard's, now a tarmac way known as the Innocent Railway Path, and an even shorter stretch in the town of Dalkeith itself.

ST LEONARD'S TO DUDDINGSTON PARK ROAD 2 MILES

From Dalkeith Road (the A7), go along Holyrood Park Road by the Commonwealth Pool and Pollock Halls, both situated against the stunning backdrop of Arthur's Seat and Salisbury Crags. Turn left into East Parkside. Pass on the right the entrance to St Leonard's Tunnel and continue through an area of high-density housing built on the 8-acre site of the former station and coal depot. Pass by on the right the 1835 Category B listed former 'locomotive warehouse' – 'engine shed' and other standard railway terminology took time to become established. It now houses the Holyrood Park Distillery – the first single malt distillery in the heart of Edinburgh for nearly 100 years. Beyond, by St Leonard's Lane, is St Leonard's police station, made famous by Ian Rankin's fictional detective John Rebus.

Return to the line of the railway, spanned by pedestrian walkways, and so into the lit tunnel – the start of the Innocent Railway Path. This very impressive bore, 20 ft wide and 15 ft high, is still almost bone dry. If cycling, freewheeling on the long descent is great fun, regardless of one's age. Do make as much echo noise as possible! And so into 'innocent' countryside, albeit with incongruous lamp posts. The eastern end of the tunnel is overlooked by a formation fancifully known as Samson's Ribs – a good example of columnar jointing, often formed in basic lava flows, sills and dykes by ponding and contraction during cooling. Such columns are normally hexagonal and perpendicular to the cooling surface.

St Leonard's tunnel, where the line dropped steeply towards Duddingston. *Robin Howie*

Also at the tunnel exit, but on the right, are small ponds, all that is left of the Bonnie Wells o'Wearie, disrupted during construction of the railway. Earlier maps show several troughs, pumps and wells. The waters, said to have supernatural properties 'to cure the weary traveller', made an excellent supply for local residents. In earlier times it was a site where lovers came to relax and dream, hence the beautiful traditional song.

The tarmac path continues, with Prestonfield Golf Course on the right and Duddingston Loch and the Bawsinch Nature Reserve on the left. Managed and part-owned by the Scottish Wildlife Trust, this natural freshwater loch is an important site for breeding and wintering wildfowl.

A long, straight stretch, with walls on either side, leads to the old railway bridge over the Braid Burn. Built in 1831, it employed cast iron beams made by the Shotts Iron Company. These originally extended over the full width of the railway, and are of outstanding interest as being amongst the earliest surviving examples of their type anywhere. (The similar cast iron beams of a former footbridge in Linlithgow, dating from 1842, are in the keeping of the Scottish Railway Preservation Society.)

The Innocent Railway Path. *Robin Howie*

Close by is a level crossing gate at Duddingston Road West. Walkers today cross safely by the 'green man' signals. The path then runs parallel with 'The Sub' (which endures as Edinburgh's south-side relief line) and brings the walker to Jewel Park.

Return to the city by bus – from the north side of the Park – is a ready option. Nevertheless, even with the climb back to St Leonard's, the return walk is easy.

For those who chose to add the short section from Dalkeith back to the Glenesk viaduct, the express Edinburgh-Dalkeith bus takes about 30 minutes. By train to Eskbank station, with bus connection to Dalkeith, takes about 45 minutes.

Bridge over the Braid Burn. *Robin Howie*

DALKEITH TO GLENESK JUNCTION ½ MILE

The extension into Dalkeith terminated on the west side of Eskbank Road (A6094). This station, completely rebuilt c 1870, was closed in 1942, though freight traffic continued until 1964. It was eventually demolished. The bus station on the same site has now been replaced by a supermarket with car park – a familiar tale. The station buildings and goods yard are altogether obliterated.

Head to the end of the car park, then by the A6094, very briefly, pass a car wash to reach Cemetery Road. This crosses the much filled-in trackbed by an attractive green-painted metal bridge. Make the short detour to the tall, octagonal brick building with a louvered wooden top – a water tower, but not of railway provenance. The tower was built in 1879 to hold the town's supply and its tank stored 12,000 gallons of water.

Return to the bridge and the start of the trackbed/tarmac path which gently descends, curving through a cutting with retaining walls. Giving views down the steep wooded slopes to the North Esk, the trackbed later approaches the restored and fenced off Borders Railway, across the impressive B-listed Glenesk Viaduct. Few walkers or cyclists seem to begrudge the curtailments which railway restoration brought.

It is possible to descend sharply to the right for a closer look at the viaduct – a single span structure some 60 ft above the river, with masonry abutments believed to be founded directly on rock. There are adjoining masonry wing walls on both river banks.

Dalkeith station and yard in latter-day condition. A passenger service, branching from the Waverley Route at Glenesk Junction, lasted until 1942. *Sanderson collection, courtesy NBRSG*

No evidence remains of Glenesk Junction Station, closed in 1886, which was situated north of the viaduct some two miles from Dalkeith Station.

Glenesk viaduct, spanning the North Esk river. Despite some debate, expert opinion holds that it dates back to the pioneering Edinburgh & Dalkeith Railway. *Robin Howie*

WALK 12

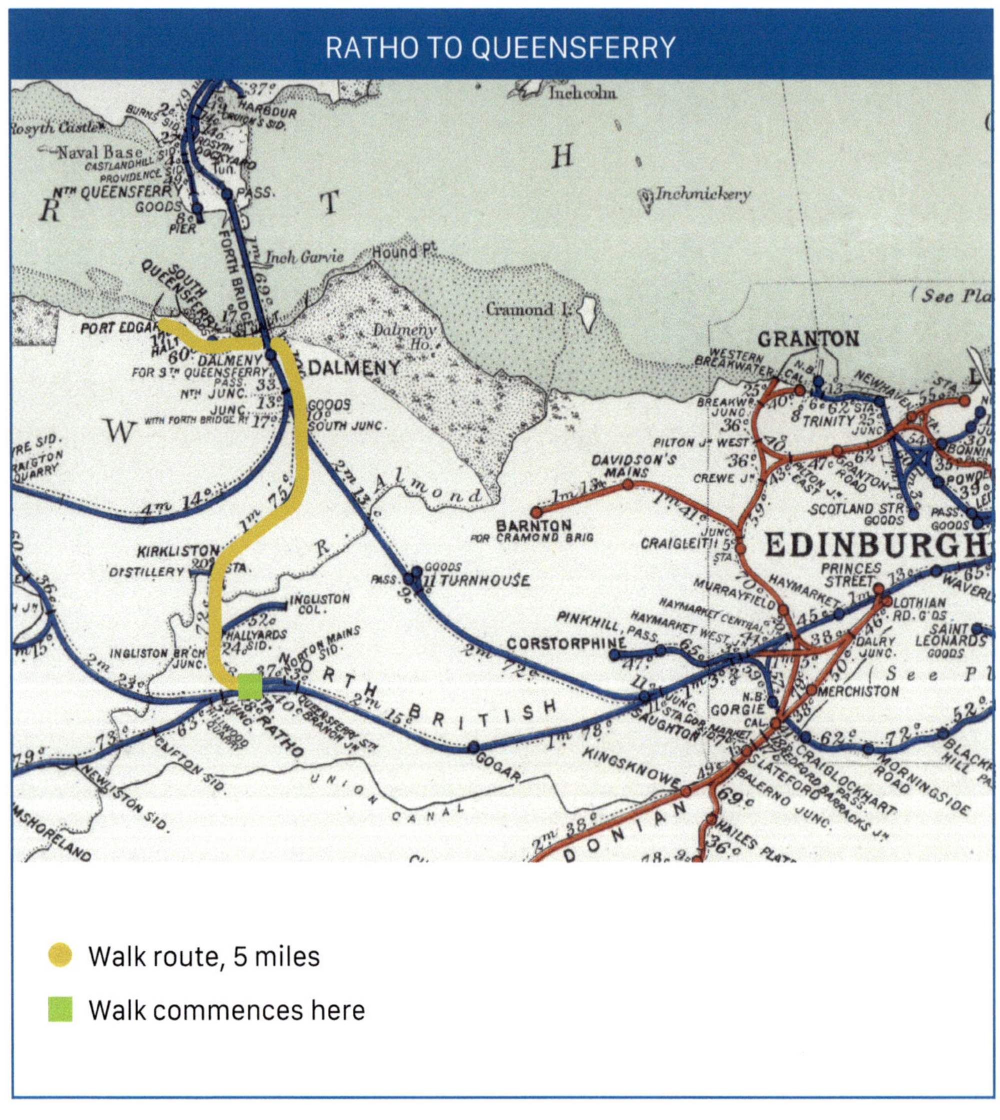

● Walk route, 5 miles

■ Walk commences here

The 5¼-mile single-track branch left the Edinburgh–Glasgow main-line at what from 1866 became Queensferry Junction, just east of Ratho station, with Bathgate Junction a little to the west. Ratho thus acquired a supplementary platform, treated as Ratho Low Level, with the original main-line station designated High Level. (They were subsequently mapped together as Ratho Station.) The branch, which dipped behind the station buildings on the up side, was opened to Dalmeny in 1866, extended to Queensferry in 1868 and further extended west in 1878 to a new station and breakwater at Port Edgar. There was also a short-lived Queensferry Halt near Jacob's Ladder, just above the former Seals Craig Hotel.

Following the opening of the Forth Bridge in 1890, passenger services on the branch terminated at Dalmeny or continued to Dunfermline. (However, trains ran again to Port Edgar, serving Butlaw Naval Hospital, during the First World War.) By contrast, freight endured. Indeed the Queensferry goods outlived other local workings of its kind. Latterly this duty regularly fell to the ex-North British engine *Maud* (of a numerous class designed in the 1880s and later rebuilt), which was finally retired in 1966 and purchased by the Scottish Railway Preservation Society. All traffic to Queensferry finally ended that same year, but not before another ex-North British veteran locomotive, the preserved *Glen Douglas* (now resident in Glasgow Transport Museum), had paid at least one visit.

Ratho was the scene of a rail crash on 3 January 1917 when an Edinburgh-to-Glasgow express collided with the branch engine. The express was packed with passengers and soldiers returning from the New Year holidays; 12 people died and 46 more were seriously injured. The soldiers immediately set about rescue and first aid, and their prompt action no doubt reduced the death toll. A collision between a main-line express and a shunting goods engine had occurred in 1911 at almost the same spot, while in yet another accident near Ratho, in 1874, Frederick Dundas, son of the 26th Laird of Dundas, had been fatally injured. The Laird, who at that date owned Inchgarvie Island (below), was a talented engineer; he established the Paragon foundry and factory for machinery on the Dundas estate.

The main-line Ratho station closed to passengers in 1951, thus denying today's walkers an arrival by train.('Ratho Station', however, remains a recognised place, cf. 'Balquidder Station' on the long-closed eastern half of the Callander & Oban Railway.) Use Lothian Buses No. 20 to reach Station Road off the A8. However, Lothian Buses, number 20, gives access from Edinburgh to Station Road. The 1975 Local Government (Scotland) Act reorganised local authority boundaries, ending Queensferry's Royal Burgh status. Formerly in the county of West Lothian, it is now administratively within the City of Edinburgh, albeit on the extreme north-western edge.

RATHO TO KIRKLISTON STATION 1¾ MILES

The line traverses undulating terrain and thus features a series of attractive cuttings – yet there are intermittent excellent views back to the Pentlands and ahead to the Forth.

Leave the A8 at the sign 'Ratho Station' and go to the top end of Station Road, then turn left to the site of the Ratho Low Level, now occupied by McNee Coach Hire. On the north side, hidden by trees, is a cutting – the start of the branch line. Return to Station Road where the far-side cutting has been in-filled and is now occupied by houses. The near side is also likely to be in-filled to consolidate the overbridge – a good viewpoint from which to admire the tips of the white 'sails' of the new Queensferry Crossing below. Descend Station Road and pass by Hillwood Primary School and the entrance to Lochend Industrial Site beyond which all trace of the trackbed curving to the modern Newbridge roundabout has been lost.

For anyone who suffers from hyperacusis, a hearing disorder characterised by intolerance to usual environmental sound, the short walk to the roundabout may be a nightmare – surely the noisiest start to any expedition of this kind. Briefly turn right at the foot of Station Road and use the pedestrian bridge to cross the very busy A8 dual carriageway. On the far side is the vast expanse of Edinburgh Airport. Pre-Covid passenger numbers exceeded 14 million per year and departing planes averaged 110 every day. The flight paths can be observed – they most certainly can be heard!

Follow the pavement by the A8 to reach the north-east side of the massive roundabout. Keep walking to another pedestrian bridge. Do not cross! Continue for a few yards and then

On the trackbed at last – beyond Newbridge roundabout. *Robin Howie*

Ex-North British veteran 'Maude' (British Railways 65243), running tender-first, takes the Queensferry Goods across the River Almond, c.1960.* *Sellar*

Kirkliston station and yard, Queensferry branch. *NBRSG*

---

* 'Conscripted' to France along with others of the same class during the First World War, this locomotive was named for General Frederick Maude. It was subsequently restored to North British condition by the Scottish Railway Preservation Society.

turn right under the bridge to find salvation – the start of a tree-fringed trackbed with a lovely central path.

The M9 curves left, but the trackbed heads straight north for one mile to Kirkliston with the noise level gradually abating. Very soon Kirkliston's oldest surviving building comes into view, the well-worth-a-visit Kirk, prominently positioned. Cross the River Almond, still spanned by the railway bridge, from where a tarmac path leads to Auldgate; here the latter-day houses on the right cover the site of Kirkliston station. There is still a Station Terrace, the sometime access, and a Station Road, the B800. To the left, on higher ground, are two more houses; the first is of modern design, the second of railway provenance. In an architectural style typical of the North British Company, it was the station master's abode and the only evidence of the town's former station. The town's Newliston Arms was established in 1865, perhaps in anticipation of the business which the branch would bring.

KIRKLISTON TO QUEENSFERRY STATION 3½ MILES

Pass under the bridge on the east side of the town, after which trackbed and minor road run parallel for over one mile to Carlowrie Cottages – another attractive stretch, initially by a modest cutting, with the minor road heard but not seen. There is one spot which may be wet. Gradually views of the Pentlands emerge to the south.

The grit path leads on between tall pine trees, then a gradual curving rise takes the trackbed across two roads. Now heading north, at first within a deep, dry and well-wooded cutting, it is a steady ascent to the line's modest summit at some 150 ft. Pass under a high arched bridge, beyond which, on the left, is a fenced-off industrial estate. The trackbed ahead is intersected by the main line to the Forth Bridge, but paths on either side lead down to a road. On the right is an oil storage depot. Crude oil from the Forties pipeline underwent stabilisation, gas-processing and treatment at Grangemouth before being transferred to the depot, whence diesel was pumped to sea-going tankers, at Hound Point marine terminal just east of the Forth Bridge; but the ageing Grangemouth facility is now out of use. From there diesel is pumped to the Hound Point marine terminal, just east of the Forth Bridge, and loaded onto tankers.

Cross to the west side of the road and use steps on the right to follow (approximately) the direction of the severed branch. Pass under the unattractive concrete bridge that carries the A90. A wider area indicates the site of the first Dalmeny station, where a goods siding was added post-1890. It is no longer obvious how the original Queensferry branch and later main line first merged then diverged. (Do not follow the sign for *today's* Dalmeny station!) Continue by the branch trackbed, now a broad tarmac way, which curves to descend some 120 ft towards the Forth.

Leaving Kirkliston. *Robin Howie*

The sole wet area… *Robin Howie*

Pass under the very high viaduct that carries the main line to the Forth Bridge *proper* – the lattice girder approach spans, south and north, are generally regarded, save by engineering purists, as integral to the Bridge, which can be viewed by climbing to the lip of the cutting but is better seen in full glory from a vantage point overlooking Queensferry.

Having already made his reputation as the author of the Forth and Tay train ferries and as a purveyor of economically constructed railways, Thomas

Bouch was was knighted for the design of the first Tay Bridge. Its collapse (28 December 1879) – the biggest civil engineering disaster in British history – ended his career. Although the North British Company contemplated make-shift repairs, and even thought of employing Bouch again, public opinion ruled otherwise and work on the discredited engineer's ambitious suspension structure at the Queensferry Narrows ceased. 1879 is embedded in Scotland's folk-memory, perhaps in consequence of William McGonigal's well-known verses. The circumstances of the disaster and Bouch's career have been the subject of repeated studies and popular accounts. His reputation has been in some measure rehabilitated and his place in civil engineering history more objectively re-evaluated.

The Forth Bridge still stands proud today, after more than 130 years. The brain child of Sir John Fowler and Benjamin Baker, it was built by contractor William Arrol's Glasgow-based company, Tancred & Arrol. The new cantilever design insured heavily – even over-insured – against the vagaries of wind and weather. Already a knight – his distinguished career stretched back to the beginnings of London's Underground – Fowler became a baronet, while Baker and Arrol were knighted.

By the site of old Dalmeny station. *Rhona Fraser*

Queensferry branch: an ex-North British Locomotive, its tender lettered 'LNER', takes a brake van down the stiff gradient from Dalmeny Junction to the Forth littoral, where the Queensferry branch swung west, under the southern approach viaduct to the Forth Bridge. *NBRSG*

Baker was assisted by Allan Stewart whose expert calculations had previously aided Bouch. Stewart, described as 'Chief Assistant Engineer', gained a first-class degree in mathematics from Cambridge University in 1853.

Every 19th-century proposal for a bridge at Queensferry Narrows looked to exploit Inchgarvie Island and the foundations for the mid-stream towers of Bouch's back-to-back suspension bridges would have been laid there. Instead the island holds the central cantilever of the Fowler-and-Baker Forth Bridge. It also provided workers' accommodation, within the reroofed castle buildings. Ironically, a solitary brick pier, the sole remnant of Bouch's abandoned work, later became the base of a lighthouse.

The German civil engineer, Wilhelm Westhofen was Assistant Engineer with responsibility for foundations and pier building and for the construction of the Inchgarvie cantilever. He wrote an invaluable and comprehensive account of the design and construction of the Bridge in which he refers to the design as 'cantilever and central girder'. (Bouch would have added what he termed 'stiffening girders' to his suspension structure,but that concept is very different.) Confusion arises in that, strictly speaking, the Bridge, as distinct from its approach viaducts, consists of three diamond trusses (towers), embodying the *cantilever principle* (see also below), while in popular and even professional usage the towers are commonly

The Forth Bridge, seen from the former railway. *Robin Howie*

described as *'cantilevers'*. The central girder, varying in depth, plan and elevation, runs continuously through the towers, named 'Queensferry', 'Inchgarvie' and 'Fife'. The Inchgarvie tower is self-supporting and a true balanced cantilever; hence it is significantly longer than the other two. It also includes additional steel and is heavier in construction. All this was to counteract the slightest risk of tipping under the weight of

Bouch's 'memorial' lighthouse. *Tom Wright*

trains, where the cantilevers met. (Short spans linked their extreme arms.)

In end-on elevation, all three towers taper, because stability demands that the base be wider than the top. In side-on elevation, viewed from a distance, the massive structure exhibits a curiously delicate tracery of girders, and it can be better appreciated how the tubular members are in compression and the lattice members in tension. The Queensferry and Fife towers are counterweighted at their meetings with the approach viaducts, which resists deflection at their meetings with the Inchgarvie tower.

Constructed over 7 years, the 1½-mile-long Bridge, 150 ft above the Forth, is one of the world's most recognised and popular engineering wonders and has been, since 2015, a

UNESCO World Heritage Site. It takes over 200,000 litres of paint to cover the bridge's 145 acres of surface, so vast that a mere shower of rain adds many tons of weight; yet, contrary to popular belief, painting was never continuous and the glass flake epoxy coating, which bonds with metal, applied in 2011, will last 25 years. What a change from days of yore!

The 'throw a penny from the train' superstition harks back to the days of steam and drop-windows. As a young lad I never queried why this practice would bring luck. A proportion of the coins lodged in the troughs between the tracks (there are no cross-sleepers) and the surfacemen reckoned on gathering and sharing £1 every week (once a very worthwhile sum).

The North British banned 'ladies' in their hazardous pre-1914 apparel from the bridge – mid-Victorian crinolines and turn-of-the-century hobble skirts were the cause of many accidents more-or-less serious, not infrequently in running for the train. By contrast Network Rail, owners of the bridge today, has submitted a full planning application to install a bridge walk to encourage visitors on a regular basis, focussing on a 'bridge climb' experience. They expect to attract 80,000 persons a year. Access will be provided via an existing walkway under the south approach spans and a new steel walkway positioned discreetly within the top member of the southern suspended span. Visitors, secured by harness and safety line, will be 'pulsed' in groups of up to 15 from a new single-storey reception hub on the South Queensferry side. But this ambitious project has made only slow progress.

Return to the trackbed/tarmac way.

Steps on the right, known as Jacob's Ladder, lead down to Hawes Pier and Hawes Inn. The first paragraph of Scott's *The Antiquary* sets the scene for an encounter at South Queensferry. 'It was early on a fine summer's day, near the end of the 18th century, when a young man, of genteel appearance, journeying towards the north-east of Scotland, provided himself with a ticket in one of those public carriages which travel between Edinburgh and the Queensferry, at which place … there is a passage-boat for crossing the Firth of Forth'; and it was at the ancient hostelry, the Hawes Inn, that the two travellers, Jonathan Oldbuck and young Loval, began their friendship and decided to continue together to the end of their journey.

In 1071 Malcolm III granted free passage at the 'Queen's Ferry' for pilgrims on their way to St Andrews. The ferry service acquired by the North British Company in the 1860s ran until 1964 when the Forth Road Bridge opened – at the time the UK's longest suspension bridge and the longest outside the USA.

Pass under a stylish skew-arch accommodation bridge, then via a broad, deep cutting on a steady descent, popular with residents and cyclists, and overlooking the coastal road. A broadening grassy area, just before the bridge on the Loan, is the site of Queensferry Station. Beyond the steep Loan is the site of the goods depot, which serviced the Vat 69 bottling and blending plant. The building was demolished in 1985 and the land is now covered by a large shopping area and car park, the fate of many a railway yard. Beyond, all trace of the line that continued to Port Edgar has been utterly subsumed by an industrial estate and modern houses.

Now there is a third bridge, officially opened on 4 September 2017 – the cable-stayed Queensferry Crossing. Though slender in appearance by comparison with the work of Fowler and Baker, it was built, like the railway bridge, *on the cantilever principle*. Strictly speaking, cantilever (or balanced cantilever) is *a method of construction,* applicable only whilst a bridge is being built, although the term is often used to describe the finished structure. The 'Queensferry Crossing: Vision to Reality' – the official account of the technical challenges of this magnificent bridge – has a superb picture,' of the 644-metre-long Central Tower span, recognised by *The Guinness Book of World Records* as the longest, free-standing balanced cantilever structure ever built – but only for some two weeks until it was joined to the other spans. Situated in mid estuary, the Beamer Rock (a dolerite intrusion) made an ideal location for the Central Tower's foundations.

Where else in the world will you find three renowned bridges of contrasting appearance, belonging to three successive centuries, to mark the end of a varied and interesting walk?

Head for Dalmeny station. From there, why not take the train over the Forth Bridge to North Queensferry, then return by rail to Edinburgh? North Queensferry offers a different, more 'close-up', view of the bridge, and the well-worth-a-visit Deep Sea World marine experience, located almost beneath the landward cantilever arm, has a most informative plaque about its construction.

The steady descent to Queensferry. *Robin Howie*

WALK 13

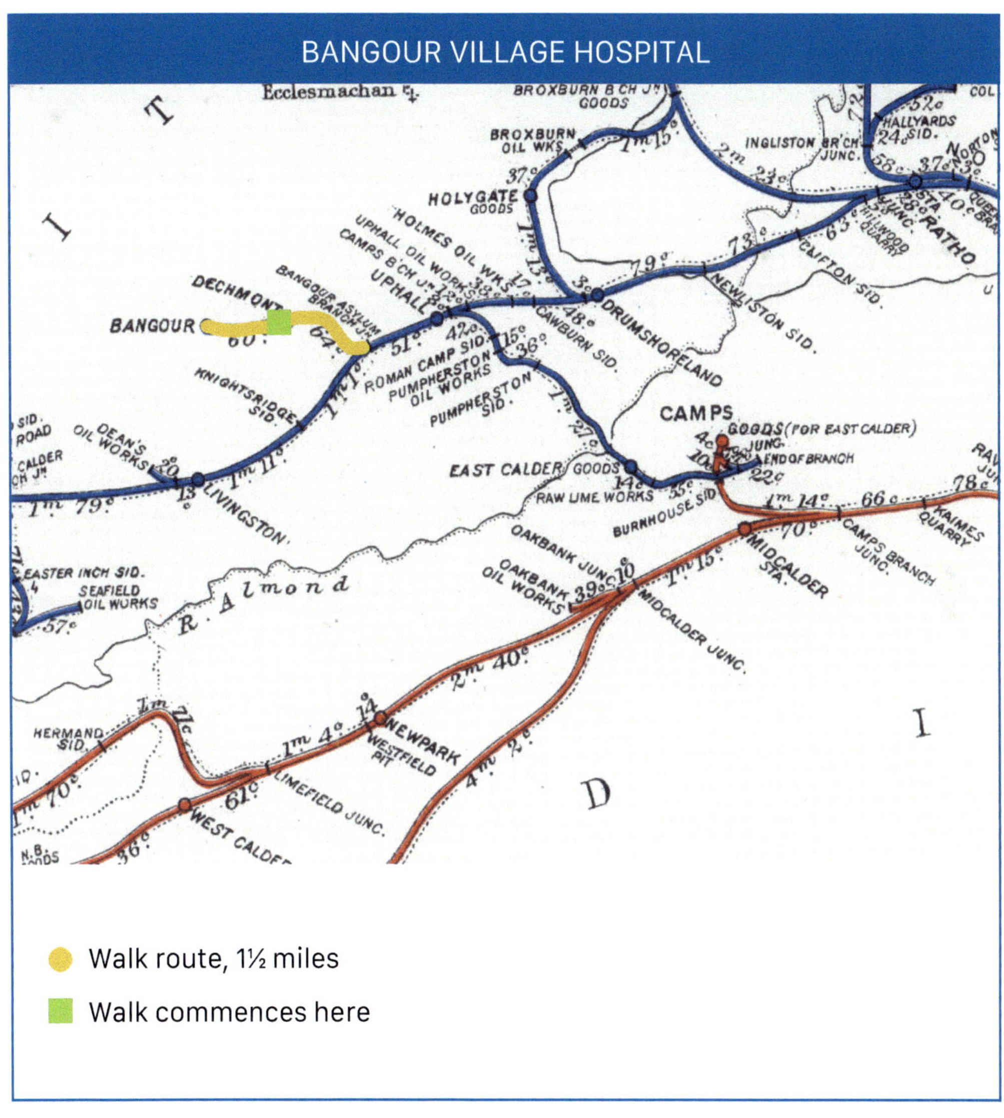

Walk route, 1½ miles

Walk commences here

It would be hard to imagine a railway that packed so much history into its short life – the passenger service operated for only 16 years (1905–1921) on this 1½ mile, single track branch line. Properly the Bangour Asylum Railway, it became affectionately known as the 'Wee Bangour' and its train as the 'Wee Bangour Express'. Curvature was tight and the speed limit 25 mph.

The Edinburgh District Lunacy Board purchased the 960-acre estate of Bangour – pronounced 'Bangower' and meaning 'the hill of wild goats' – as the site for a new asylum to provide treatment and care for 'the pauper lunatics' of the City of Edinburgh. Patients would benefit from the peaceful rural setting, some 14 miles from the city, and might be readily occupied labouring on the land. At a time when road haulage was in its infancy, the railway was originally planned to carry the building materials for the construction of Bangour Village (as the asylum became known); it later provided the only easy means of access for inmates and the supplies needed to care for them, besides staff and visitors.

The North British Railway entered into a working and maintenance agreement with the Board, who committed to laying the track, paying all overheads and dismantling the railway if it ceased operation. The North British were assured of 50% of the gross revenue and an annual minimum of £1,500, with the Board bound to find any deficit – these were not unusual terms but neither were they generous*. (The branch never paid its way.)

The branch left the Edinburgh–Bathgate route at a junction west of Uphall and ran to an intermediate station at Dechmont, the small village which housed many of the hospital employees. This section was operated as a public railway, whereas the terminus at Bangour was shown in the public timetable as Bangour (Private). The first passenger train ran on 19 June 1905; there was no special ceremony. Guests arrived by rail for the official opening of the Village in October 1906, as guests did again in June 1908 when the Village Hospital was opened. Both occasions are commemorated by a plaque in the administration block of the Hospital.

During the hostilities of 1914-18, Bangour Village was converted into the Edinburgh War Hospital – the largest such facility in Scotland. Psychiatric patients were evacuated to other asylums by special hospital trains.

From June 1915 casualties, evacuated from the Western Front to the docks at Southampton, were transferred by hospital trains direct to Bangour, a 12-14 hour journey. The installation of a tented hospital for those less severely wounded or otherwise incapacitated increased potential patient numbers to over 3,000.

Following the Armistice, these numbers gradually reduced and the hospital closed in December 1921. Bangour Village reopened as an asylum, with the first of the returning patients arriving in August 1922 – and by motor bus, a harbinger of the little railway's fate.

---

* At the turn of the century railway companies in general experienced a stiff increase in working expenses.

'The Wee Bangour Express' — a three-coach train and locomotive. These diminutive tank engines did local duty all over the North British system, from Jedburgh to Fort Augustus. Hennigan collection, courtesy Lynn.

With ever improving road access, the need for a peace-time railway quickly diminished, and the Board decided that it would be too costly to continue. (Even before 1914 the Board had had to find around £1,000 every year, because half the revenue did not cover the minimum.) Passenger services ceased in May 1921, freight three months later, and the track was lifted thereafter.

September 1939 saw the Village Hospital once again transformed into the Edinburgh War Hospital. To increase capacity, an Annexe (long, low prefabricated buildings) was sited outside the grounds as an Emergency Services Hospital. By spring 1941 the first patients were admitted, none of them blitz victims or war casualties, and Bangour gradually became a combined civilian and service facility, later known as Bangour General Hospital. In fact, this reprieve extended some 35 years more, not the only example of war-time provision giving long service under the NHS. However, in 1990 Bangour was replaced by West Lothian's St John's Hospital in Livingston, after which Bangour Village was closed.

My wife, Dr Margaret Cook, was a consultant haematologist at Bangour General Hospital and our 2020 visit to both Village and General Hospital was a nostalgic trip. The Village is now a sad place with many of the abandoned buildings awaiting demolition. Even the large and austere Memorial Church (completed and officially opened in 1929 to commemorate Bangour's wartime role) is no longer used; as for the now demolished General Hospital, only the site can be identified.

Ambulance train and an injured soldier at Bangour station. *West Lothian Council*

At what will remain the main entrance off the A89, a notice proudly proclaims Bangour Village to be 'a residential and mixed-use site, the natural choice for the perfect home'. It will become a mixture of new-build housing, renovation, and listed buildings converted to modern use. Construction has started with the formation of roundabouts, access works, a cycleway, footpaths, car parking and landscaping.

The M8 together with the Dechmont bypass (A89) have largely obliterated any trace of the public section of the railway to Dechmont, whereas some of the Dechmont–Bangour stretch, the private section, remains.

An out-and-back 3-mile walk starts from Dechmont, in the direction of sometime Bangour Junction. At the foot of Burnside (a short road off Main Street) there was once a level crossing. Do not go along the broad drive straight ahead – this was Mortuary Road, still so called today – instead, turn sharp left on the narrow lane, known locally as the Railway Path. The ballast underfoot tells of its railway origins. Becoming grassy (and narrow during high summer growth), it leads between houses on the left and a wooded area on the right. When the area on the left opens out to a view of the A89, the ½ mile walk to the site of Bangour station is over all too soon. At the far end, on the right, are remnants of the brick-built platform. No trace remains of the timber-clad station building. Beyond this point the route is currently blocked by construction fencing. The line originally extended west to the hospital transport depot beside the road entrance from the A89. There were three sidings, one of which allowed direct transfer of coal to the power station boiler house, which produced steam

for the laundry, kitchen and bakery as well as generating electricity.

Return to Burnside.

Follow the line of the railway via Goodall Place, then by a short path on the left to Goodall Crescent, which curves right to Main Street and the 18-feet-deep Dechmont Cutting, now filled in by spoil from road improvements. This spot is identified by a slight rise in Main Street and what remains of the roadside parapets. Houses on the far side cover the site of single-platform Dechmont station and the line of the railway. Note the garden planters made from old railway sleepers.

Cross Main Street to Knightsridge Road and the signed Loan Path. At the end of the

Left: The Memorial Church. *Robin Howie*
Below: The Church Plaque. *Robin Howie*

The Railway Path. *Robin Howie*

Trees are taking over the remains of Bangour station platform. *Robin Howie*

The site of Dechmont cutting. *Robin Howie*

road, use the traffic island to cross the very busy A89 and regain the trackbed, in fact a straight and broad tarmac way (still the Loan Path), which bisects the North Wood (Woodland Trust for Scotland). The Path leads over a spur of the M8 by way of an ugly concrete bridge and then to a concrete pedestrian underpass beneath the motorway. The straight tarmac way, now lit, continues to follow the line of the railway (though no evidence of that remains) and rises to a road. To the right is Deer Park, while to the left is Deerways Business Park. From this point the railway curved left to reach the Edinburgh-Bathgate-Airdrie line. Continue on the tarmac way (no longer coterminous with the lost branch) to a footbridge over the now-electrified Bathgate tracks. Turn left and look in vain for any sign of Bangour Junction, which was located where the now-widened A899 crosses the railway.

It is a short return to Dechmont by the same route.

WALK 14

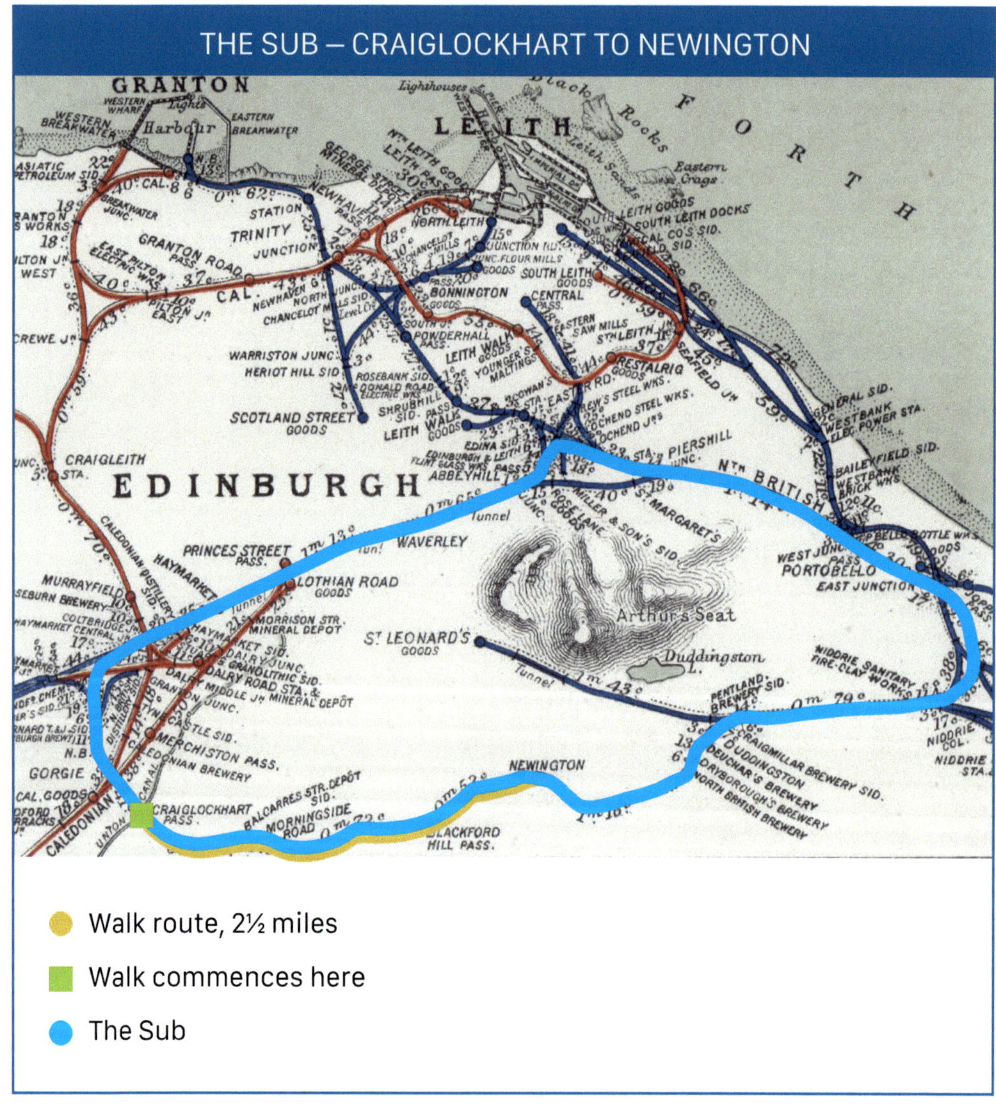

Walk route, 2½ miles

Walk commences here

The Sub

This expedition is 'different' in that The Sub remains a working railway and certainly should not be 'walked'! It has belonged to the wider Edinburgh townscape for more than 140 years.

A line across the slopes rising towards the Pentlands, by-passing the then built-up area, obtained Parliament's approval in 1865, at which date the reconstituted North British Company of 1862, now united with the Edinburgh, Perth & Dundee, was about to absorb the Edinburgh & Glasgow. The scheme languished but was successfully revived some fifteen years later as the Edinburgh Suburban & Southside Junction Railway, from Duddingston, on the sometime Edinburgh & Dalkeith Railway, to the triangle formed by Gorgie Junction, Haymarket Central Junction and Haymarket West Junction. This nominally independent enterprise existed as the client of the North British, who for the moment entered into a working and maintenance agreement. Almost half the shares were taken by financiers associated with the working company – a device which the North British favoured – and all the shareholders earned a 5% premium when eventually bought out.

The engineer was Thomas Bouch, designer of the first railway viaduct across the Firth of Tay, which was to collapse under a train in a winter gale on 28 December 1879 – the worst railway disaster that Britain had so far experienced. Bouch estimated the cost of his by-pass line at £250,000, and – notwithstanding all the drama on the Tay, which effectively ended his career – the project went forward in that very parliamentary session.

Opening for goods traffic came in October 1884. Passenger traffic began two months later, for which an island platform (with its own booking office) was added to the disjointed Waverley Station of the time. (The present detached 'island', below and parallel with Market Street, dates from Waverley's thorough remodelling in the 1890s.) When new, the station buildings at Duddingston, Newington, Blackford Hill, Morningside Road, Craiglockhart and Gorgie earned modest praise, as superior to the generally dowdy North British article.

Serving the city's expanding southern districts, the new railway became an Edinburgh institution, soon affectionately known as 'The Sub'. From the early 20th century, the standard timetable was modified to embrace Leith Central[22]. Locomotives on local duty were for the most part stabled at St Margaret's – not Haymarket, where main-line engines predominated.

From the 1930s the latest LNER motive power worked alongside older North British types. Late in the day, DMUs were briefly employed. Meanwhile, British Railways had added a useful spur from ex-North British Craiglockhart to ex-Caledonian Slateford, which enhanced the value of 'The Sub' as a relief line for long-distance freight[23]. During the last months of Princes Street Station (closed in 1965), with the new Duff Street link into Haymarket (and thence Waverley) incomplete, Sunday trains for Carlisle and the South via Carstairs left from Waverley's east end for a half-hour detour by the southern side of the city,

---

22 See 'Edinburgh's railways – a brief history'
23 The General Map (p.1), of much earlier date, does not include the Slateford spur, but its location is obvious.

EDINBURGH SUBURBAN and SOUTHSIDE JUNCTION RAILWAY and ST LEONARDS BRANCH. July 1896.
Inner Line Circle Trains—*continued*. *Week-Days*.

Stations and Sidings	51 1st Sighthill and North Leith, Exp. Goods	52 Empty Coaching Stock, From Perth	53 St Leonards, Min.	54 Suburban, Goods	55 Pass. 1 & 3 Class	56	57 Goods	58 First Sighthill and Berwick, Goods ex. Mon.
—Edinburgh (Waverley) dep.	From Sighthill	From Perth	…	…	1 45	…	…	From Sighthill
—Haymarket Station	…	…	…	1 30	1 49	…	…	…
—Haymarket Central Junc.	…	…	…	1 31	1 50	…	…	…
—Haymarket West Jn.	1 3	1 19	…	…	…	…	…	…
—Gorgie	1 5	1 21	…	1 40	1 52	…	…	2 0
—Craiglockhart	…	…	…	…	1 55	…	…	…
—Morningside Road	1 10	1 25	…	1 50	1 53	…	…	…
—Blackford Hill	1 14	1 27	…	2 8	2 1	…	…	2 15
—Newington	…	…	…	…	2 3	…	…	…
St Leonards	…	…	1 30	…	…	…	…	…
—Duddingston	1 18	1 31	1 50	2 15	2 7	…	…	2 25
—Niddrie West Junc. arr.	1 25	…	1 55	2 20	…	…	…	2 30
—Do. Do. dep.	1 35	1 36	…	2 35	…	…	2 48	2 55
Niddrie South Junc.	…	…	Stop.	…	…	…	…	…
—Niddrie East Junc.	…	…	…	…	…	…	…	4 59
—Niddrie North Junc. dep.	1 40	1 41	…	…	2 11	…	…	…
—Portobello arr	…	1 44	…	2 40	2 13	…	…	…
South Leith	…	…	…	…	2 58	…	…	…
—Portobello dep.	1 44	Stop.	…	Stop.	2 14	…	…	…
—Pierhill Junction	1 48	…	…	…	2 17	…	…	…
Pierhill	…	…	…	…	2 18	…	…	…
—Lochend Junction	…	…	…	…	2 19	…	…	…
—Abbeyhill	…	…	…	…	2 22	…	…	…
Leith Walk arr.	…	…	…	…	…	…	3 20	…
North Leith	2 5	…	…	…	2 45	…	3 45	…
Granton	…	…	…	…	…	…	3 30	…
—St Margarets dep.	Stop.	…	…	…	…	…	…	…
—Abbeyhill Junction	…	…	…	…	2 23	…	…	…
—Edinburgh (Waverley) arr.	…	…	…	…	2 26	…	…	…

Stations and Sidings	59 Haymarket and St Boswells, Goods	60 Pass. 1 & 3 Class	61 Niddrie and Perth, Thro' Goods	62	63 Second Sighthill and Berwick, Exp. Goods	64	65 Pass. 1 & 3 Class	66
—Edinburgh (Waverley) dep.	…	2 45	From Perth	…	From Sighthill	…	3 45	…
—Haymarket Station	2 30	2 49	…	…	…	…	3 49	…
—Haymarket Central Junc.	2 34	2 50	…	…	…	…	3 50	…
—Haymarket West Jn.	…	…	…	…	…	…	…	…
—Gorgie	…	2 52	3 0	…	3 10	…	3 52	…
—Craiglockhart	…	2 55	…	…	…	…	3 55	…
—Morningside Road	2 38	2 58	…	…	…	…	3 58	…
—Blackford Hill	2 43	3 1	3 4	…	…	…	4 1	…
—Newington	…	3 3	3 9	…	…	…	4 3	…
St Leonards	…	…	…	…	…	…	…	…
—Duddingston	2 50	3 7	3 15	…	…	…	4 7	…
—Niddrie West Junc. arr.	2 55	…	3 20	…	3 29	…	…	…
—Do. Do. dep.	3 15	3 10	3 30	…	3 39	…	4 19	…
Niddrie South Junc.	3 17	…	…	…	…	…	…	…
—Niddrie East Junc.	…	…	…	…	3 43	…	…	…
—Niddrie North Junc. dep.	…	3 14	…	To Carlisle	…	…	4 11	…
—Portobello arr	…	…	…		…	…	4 13	…
South Leith	…	…	…		…	…	4 48	…
—Portobello dep.	…	3 15	Stop.		…	…	4 14	…
—Pierhill Junction	…	3 18	…		…	…	4 17	…
Pierhill	…	3 19	…		…	…	4 18	…
—Lochend Junction	…	3 20	…		…	…	4 21	…
—Abbeyhill	…	3 22	…		…	…	4 24	…
Leith Walk arr.	…	…	…		…	…	…	…
North Leith	…	…	…		…	…	4 47	…
Granton	…	…	…		…	…	…	…
—St Margarets dep.	…	…	…		…	…	…	…
—Abbeyhill Junction	…	3 24	…		…	…	4 25	…
—Edinburgh (Waverley) arr.	…	3 26	…		…	…	4 28	…

Stations and Sidings	67 St Leonards, Goods	68 Goods Wed. only	69 St Margarets and Moss Lye, Loco. Thro' Min.	70 Pass. 1 & 3 Class	71 Cattle Tues. only	72	73 Pass. 1 & 3 Class	74	75
—Edinburgh (Waverley) dep.	Runs only when required	4 30	From Moss Lye	4 45	…	…	5 45	…	…
—Haymarket Station		4 34		4 49	5 25	…	5 49	…	…
—Haymarket Central Junc.		…	4 44	4 50	5 27	…	5 50	…	…
—Haymarket West Jn.		…	…	…	…	…	…	…	…
—Gorgie		…	4 46	4 52	5 30	…	5 52	…	…
—Craiglockhart		…	…	4 55	…	…	5 55	…	…
—Morningside Road		4 38	4 50	4 58	5 34	…	5 58	…	…
—Blackford Hill		4 41	4 53	5 1	5 39	…	6 1	…	…
—Newington		…	…	5 3	…	…	6 3	…	…
St Leonards	4 30	…	…	…	…	…	…	…	…
—Duddingston	4 43	4 48	5 15	5 7	5 45	…	6 10	…	…
—Niddrie West Junc. arr.	4 48	4 52	5 20	…	5 50	…	…	…	…
—Do. Do. dep.	…	5 15	5 35	5 10	6 7	…	6 13	…	…
Niddrie South Junc.	Stop.	…	…	…	6 9	…	…	…	…
—Niddrie East Junc.	…	…	…	…	…	…	…	…	…
—Niddrie North Junc. dep.	…	5 18	5 37	5 11	…	…	6 15	…	…
—Portobello arr	…	5 20	…	5 13	…	…	6 16	…	…
South Leith	…	…	…	5 26	…	…	…	…	…
—Portobello dep.	…	Stop.	5 42	5 14	…	To Carlisle	6 17	…	…
—Pierhill Junction	…	…	5 44	5 17	…		6 20	…	…
Pierhill	…	…	…	5 18	…		6 21	…	…
—Lochend Junction	…	…	…	5 19	…		6 22	…	…
—Abbeyhill	…	…	…	5 22	…		6 28	…	…
Leith Walk arr.	…	…	…	…	…		…	…	…
North Leith	…	…	…	5 48	…		6 47	…	…
Granton	…	…	…	…	…		…	…	…
—St Margarets dep.	…	…	…	…	…		…	…	…
—Abbeyhill Junction	…	…	…	5 23	…		6 36	…	…
—Edinburgh (Waverley) arr.	…	…	…	5 26	…		6 32	…	…

No. 51.—Carries Unusual Despatch Signal.
No. 52.—Vehicles brought to Portobello by this Train will be taken forward by 3-0 p.m. North-Eastern Empty Coaching Stock Train from Portobello to Berwick.
No. 53.—Shunts at Duddingston and Niddrie (West) as required.
No. 54.—Carries Road Van ticketed "Haymarket to Edinburgh;" the Waybills to be delivered to Station Master, Portobello. *N.B.*—If a telegram is received at Gorgie from Bellgrove stating that there is Live Stock for Suburban Stations coming by the 1-20 p.m. Train from Sighthill, the Station Master at Gorgie will order this Train to return from Blackford Hill for the Live Stock before proceeding to Portobello. *Shunts at Blackford Hill for Passenger Train.*
No. 59.—Calls at Blackford Hill only to lift Live Stock. Takes Edinburgh Traffic to Niddrie West.
No. 63.—Carries the *Unusual Despatch Signal*. Load to be restricted to 30 Wagons from Niddrie West to Berwick. Takes Edinburgh Traffic to Niddrie West. Calls at Gorgie only to leave off Live Stock and to lift one Wagon for England.

No. 67.—Waits at St Leonards till 5-0 p.m. for Ale Traffic if necessary.
No. 68.—Engine returns immediately to Haymarket with Guard and Van only.
No. 69.—*Shunts at Duddingston for Passenger train.*
No. 71.—Takes forward to Niddrie West East Coast Traffic to go on from Portobello by 7-45 p.m. Train, Leith Walk to Berwick.
No. 73.—This Train may be kept at Haymarket for the 4-0 p.m. from Anstruther when the latter is in sight.

'The Sub', inner (anti-clockwise) circle and eastbound, with St Leonard's included – a working timetable of 1896. From 1903 some passenger services round 'The Sub' operated in and out of Leith Central and did not describe a true circle. *NBRSG reproductions*

re-joining their normal path at Slateford. Excursions and special workings, especially those powered by heritage steam, have often been similarly routed, and the sometime 'Santa Specials' round 'The Sub', organised by the Scottish Railway Preservation Society are fondly remembered.

The regular passenger service ceased from 1962. Nevertheless, restoration has been canvassed repeatedly and a shuttle service between Waverley and Newington via Haymarket may yet prove feasible.

Craiglockhart to Newington

An exploration tracing the 2½ mile section from Craiglockhart (via Morningside Road and Blackford Hill) to Newington illustrates Edinburgh's southward growth during the late 19th century. Beyond Cameron Toll the route goes round the south side of Arthur's Seat, eventually merging with the much older Edinburgh & Dalkeith Railway from St Leonard's, and cannot be so readily followed.

Just east of Meggetland, where Colinton Road parallels the Union Canal, the railway passes under both canal and road by a 65 yards-long tunnel, to emerge (on the south side) in a deep, tree-lined cutting – the site of Craiglockhart station, of which little remains.

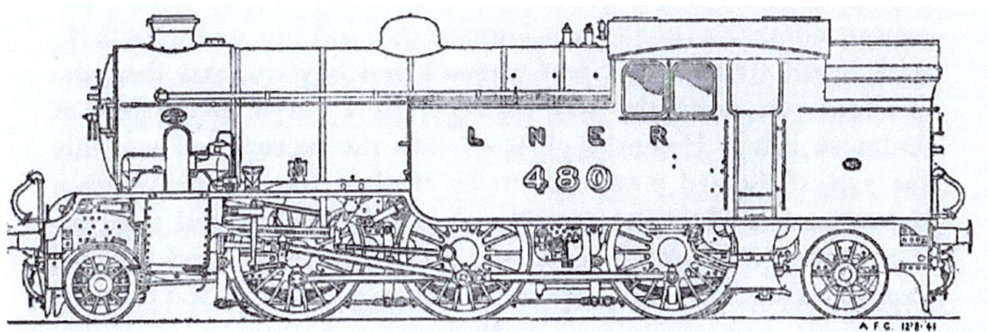

From the 1930s, new LNER tank engines appeared on Edinburgh suburban duties but did not altogether supersede earlier North British locomotives.

During excavation of the cutting, some 300,000 tons of spoil were removed. Follow a 4 feet-wide path on the north side of the railway to reach Myreside Road. Cross the road (with care) and enter the community gardens of the Royal Edinburgh Hospital, where a broad and popular path follows the southern perimeter of the grounds. (To the right, the railway remains hidden beyond a high masonry wall.) Continue by the paths which lead to the Hospital's east entrance and turn left, then right, onto Millar Crescent, which leads to Morningside Road.

Turn right, then right again, onto Maxwell Street and branch left after a modern block of flats. Straight ahead, steps lead to a lattice pattern iron footbridge, spanning the line at the western end of the former Morningside Road station. The bridge offers good views, not only of the street-level station building but also of the occasional freight train. (It is difficult to explain why such moments can be so exciting!)

Cross the bridge, turn left on Balcarres Street and left again onto Morningside Road. Opened as 'Morningside', the station was renamed 'Morningside Road' a year later. This coincided with the North British Company's formal takeover of the suburban line and avoided confusion with their other Morningside in North Lanarkshire. Best viewed from across Morningside Road, the small, slated station building, housing a few small businesses, is overlooked by tenements. A locked wooden gate closes off the steps leading to the eastbound ('inner circle') platform.

Head east on Cluny Avenue, where the unpaved path, adjacent to the railway, gives a pleasing stroll. Cross Braid Avenue (the bridge offers good views down to the railway) and continue to the end of Cluny Avenue. Follow Cluny Place to where, on the left, is an old sign, 'Astley Ainslie Hospital Boiler House Entrance'. This directs the walker to a tarmac path on the left, leading uphill. Turn right (east) to join a road. Pass by the NHS Lothian Smart Centre. (SMART stands for 'South-east Mobility and Rehabilitation Technology'.) Opened in 2007, it houses clinical, technical, office and storage facilities. Head for the high gates at the west end of South Oswald Road – if these are locked, use the pedestrian side-exit.

Morningside Road, a side-platform station with street-level entrance. From Gorgie ('circle' via Haymarket), the line climbed through Craiglockhart to Morningside, then descended through Blackford Hill to Newington. *NBRSG*

Island-platform Newington: the 'distant' signal, positioned with some ingenuity (as sighting required), is modern but otherwise the station, like others on 'The Sub', saw little change across 80 years. *NBRSG*

Follow South Oswald Road, then turn right onto Oswald Road, which dips south to cross the railway. Both the Hermitage of Braid and Blackford Hill Local Nature Reserve lie southward, overlooked by Blackford Hill. Created in Victorian times from a hollow of Ice Age origin, Blackford Pond has a small bird house in the middle and attracts swans, mallards, tufted ducks, coots and passing rarities all year round – which has made it a mecca for bird watchers from around the city. It is popular, too, with families. Do pay a visit! The pond often freezes over sufficiently to support the weight of birds but not, with any safety, the weight of humans.

Follow the broad pavement by Charterhall Road then turn left onto Blackford Avenue and so to the point where the railway passes below – this, to the left, was the site of Blackford Hill station, of which little remains. Nine extant railway stations in Scotland have 'hill' incorporated in their names, as do three former Edinburgh stations (Abbeyhill, Pinkhill and House o'Hill Halt), making Blackford Hill station (two words) something of an oddity.

Walk north past the Blackford Avenue shops and turn right onto West Relugas Road/ Relugas Road. Continue east to Mayfield Road, turning right to reach the bridge spanning the railway. Look east over the parapet towards the remains of Newington station, where the constricted site demanded an island platform.

Newington's remains today. *Robin Howie*

Head back along Mayfield Road, turning right into Mentone Terrace and Gardens. The car park of the Edinburgh People's Theatre offers a clearer view of the station. Turn right onto Mayfield Gardens for a short distance. Here twin pillars mark the locked gates at the top of the steps leading down to what is left of the platform.

From the last years of Edinburgh's suburban trains when British Railways had substituted modern second class (now standard class) for traditional third class. *Winkler*

THE FUTURE?

Lochend South Junction (Piershill loop). The tracks, seen from Marionville Road, led to Waverley (via Abbeyhill); to Granton and North Leith via Powderhall; and to Leith Central via Lochend North Junction. Today the Waverley-Portobello main line threatens to become a bottleneck. *NBRSG*

In any review of Scotland's rail revival since the 1990s, the Borders Railway (Edinburgh-Tweedbank) has pride of place. More recently, the Thornton Junction-Leven branch line has reopened (2024), while between Edinburgh and Berwick the stations at East Linton (2023) and Reston (2022) have been reinstated. A restored rail service for Winchburgh 'new town', rapidly expanding beyond the former village, has been agreed in principle. However, confidence has faltered post-Covid, and anything heroic – for example, extension of the Borders Railway to Hawick, or construction of the Almond Chord[24] – seems out of the

---

24 See 'Edinburgh's railways – a brief history'

reckoning in the current financial climate. Further electrification on a case-by-case basis, is possible, calculated to release the most-up-to-date DMU stock for other duties. Progress towards battery-powered trains is promised – witness the recent installation of overhead equipment for charging (not yet 'live'!) between Saughton Junction and Dalmeny. On the other hand, environmental concerns loom large, and there is no early prospect of a new generation of DMUs.

Edinburgh itself has seen no reopenings comparable with Glasgow's successful shuttle services to Anniesland, Springburn and Paisley Canal; but the capital's new stations at South Gyle (1985), Park (2003) and Gateway (2016) deserve mention – and the two last have tram interchange. The original tram route between Edinburgh Airport and York Place (2014) has been extended to Leith, doubling back to Newhaven; and, arguably, the now pressing question is whether this lone and wandering line, with some of the characteristics of a light railway, can develop more ambitiously. A branch by the refurbished North Bridge in the direction of Newington suggests itself. Another possibility is a branch from Roseburn, reclaiming for modern urban transport at least the Caledonian viaduct over the Water of Leith and perhaps the sometime railway to Granton in its entirety (see Walk 3: Roseburn to Granton) – the immediate uproar among walkers and cyclists opposed to this idea was considerable.

Conversion of 'The Sub' to 'light' operation or otherwise integrating it with new south-side tram routes has been proposed more than once; but this is scarcely compatible with its retention as a relief line. A shuttle between Waverley and, say, Newington is conceivable, using conventional stock, but reintroducing a 'circle' seems beyond realistic expectation. The layout east from Waverley has been pared down; the loop line by Abbeyhill and Piershill has been eliminated; and the East Coast Anglo–Scottish timetable is much more intensive than in times past. Track occupation between Waverley and Portobello is already at a premium.

STATION OPENINGS AND CLOSURES

Caledonian Railway and constituents, LMS and BR

	Opened	Closed
Balerno	1874	1943
Barnton for Cramond Brig *Cramond Brig* to 1903	1894	1951
Colinton	1874	1943
Craigleith *Craigleith for Blackhall* from 1922	1879	1962
Currie Served from Curriehill on CR main line (then styled 'Currie'), 1848–74	1874	1943
Dalry Road Previously the ticket platform for Princes Street[25]	1900	1962
Davidson's Mains *Barnton Gate* to 1903	1894	1951
East Pilton	1934	1962
Granton Gas Works	1903	1942
Granton Road	1879	1962
House o'Hill Opened as *House o'Hill Halt*, 1937–8	1937	1951
Juniper Green	1874	1943
Kingsknowe* *Slateford* to 1853, thereafter *Kings Knowes* but subsequently altered (date uncertain); *Kingsknowe Halt*, 1930–4; closed 1964, reopened 1971	1848	1964
North Leith Leith to 1903, North Leith to 1952, Leith North to 1962	1879	1962
Merchiston	1882	1965
Murrayfield	1879	1962
Newhaven	1879	1962
Ravelrig Uncertain status, intermittent use	1884	1945
Slateford See also **Kingsknowe**	1871	

---

25 Until corridor trains with gangways became the norm, special platforms for ticket collection sited on the approaches to principal stations were not uncommon.

North British Railway and constituents, LNER and BR

	Opened	Closed
Abbeyhill Sometimes *Abbey Hill* (?)	1869	1964
Balgreen Halt	1934	1968
Blackford Hill*	1884	1962
Bonnington* Originally *Bonnington Bridge* (?)	1846	1947
Corstorphine	1902	1968
Craiglockhart* Closed during Edinburgh Exhibition, 1890–1	1887	1962
Dalkeith* Originally 4 ft 6 in. gauge	1838	1942
Dalmeny On Queensferry branch	1866	1890
Dalmeny Opened as *Forth Bridge* (August–October 1890), *Dalmeny for South Queensferry* to 1919	1890	
Duddingston Intermittently *Duddingston & Craigmillar*, 1885–1954	1884	1962
Easter Road*	1891	1947
Easter Road Park Halt Alight only; last used 1964 (?)	1950	1967
Eskbank *Gallowshall* to 1850; *Eskbank & Dalkeith* from 1954	1847	1969
Fisherrow Originally 4 ft 6 in. gauge	1831	1847
Gogar	1842	1930
Gorgie *Gorgie East* from 1952	1884	1962
Granton*	1846	1925
Inveresk *Musselburgh* to 1847; *Inveresk Junction* 1876–90	1846	1964
Jock's Lodge	1847	1848
Joppa	1859	1964
Junction Bridge* *Junction Road* to 1923	1869	1947
Kirkliston	1866	1930
Leith Central	1903	1952
Leith Walk*	1868	1930
Meadowbank Intermittently used on special occasions; variously *Queen's, St Margaret's, Victoria, Palace*	1850	1906
Morningside Road Opened as *Morningside* (1884–5)	1884	1962
Musselburgh	1847	1964
Newhailes	1847	1950
Newington	1884	1962

North Leith* Because goods facilities continued, renamed *Leith Citadel* in 1952 when ex-Caledonian *North Leith* became *Leith North*	1846	1947
Piershill*	1891	1964
Pinkhill*	1902	1968
Port Edgar Restored for naval use, 1916–29 (?)	1878	1890
Portobello Superseded Edinburgh & Dalkeith station, remodelled 1890	1846	1964
Powderhall	1895	1917
Ratho *Ratho Junction*, 1851–90 (after opening of Bathgate line); additional platform for South Queensferry branch, 1866	1842	1951
Rosewell & Hawthornden *Hawthornden to 1928*	1855	1962
Saughton* *Corstorphine* to 1902; *Saughton Junction* for a period after 1890 (?)	1842	1921
Scotland Street Connected with Canal Street ('Princess Street') terminus from 1847; limited service to Granton continued briefly after 1868	1842	1868
South Leith Variously *Constitution Street, Seafield, The Shore, Leith*		
South Queensferry Relocated 1878, closed 1890, reopened as *South Queensferry Halt* 1919	1868	1929
Trinity*	1846	1925
Turnhouse	1897	1930

*Closed as wartime measure, 1917–19

FURTHER READING

George Dow *The First Railway Across the Border* LNER, 1946

Ann Glen *Edinburgh Waverley, a Novel Railway Station* Lily Publications, 2013

W. F. Hendrie and D. A. D. Macleod *The Bangour Story, a history of Bangour Village and General Hospitals* Aberdeen University Press, 1991

Harry Knox *Edinburgh, St Margaret's, 1845-1967* Lightmoor Press, 2015
Haymarket Motive Power Depot, Edinburgh, 1842-2010 Lightmoor Press, 2011
Steam Days at Haymarket Irwell Press, 2007

A.A. Maclean *The Edinburgh Suburban and Southside Junction Railway* Oakwood Press (Stenlake), 2006

Graham Priestley *The Water Mills of the Water of Leith* Water of Leith Conservation Trust, 2001

David Ross *The Caledonian, Scotland's Imperial Railway, a History* Stenlake, 2013

The North British Railway, a History Stenlake, 2014

Donald Shaw *The Balerno Branch and the Caley in Edinburgh* Oakwood Press, 1989

David Spaven *The Railway Atlas of Scotland* Birlinn/National Library of Scotland, 2015

David Watt *The Queensferry Crossing, Vision to Reality* Lily Publications, 2017

Kenneth J. Williamson *Edinburgh, Granton and Leith Railways* Amberley, 2023

A. N. Wilson *A Life of Water Scott, The Laird of Abbotsford* Pimlico, 2002

M. J. Worling *Early Railways of the Lothians* Midlothian District Libraries, 1991

Tom Wright *Glimpses of Times Past, Leith* Luath Press Ltd, 2014, reprinted 2017, 2019, 2020

A.J. Youngson The Making of Classical Edinburgh Edinburgh University Press, 1966

The back numbers of the North British Railway Study Group Journal are a valuable resource – for example, the series of articles by Donald Cattanach and Allan Rodgers constitute a deep and definitive history of Waverley Station, while the late Douglas Yuill has provided a thorough treatment (again, in successive articles) of many lines in and around Edinburgh. Back numbers of The True Line, produced by the Caledonian Railway Association, are likewise very useful.

Readers are also referred to the National Library of Scotland's map collection, in which the stage-by-stage development of both main lines and local lines in Edinburgh, the Lothians and Fife can be traced. For those particularly drawn to transport history, the complexities of company ownership and the detail of vanished junctions and connecting lines are to be found in the definitive maps and diagrams produced by the Railway Clearing House (1842-1963).